IEE Management of Technology Series

Series Editor: G.A. Montgomerie

THE BUSINESS OF ELECTRONIC PRODUCT DEVELOPMENT

Previous volumes in this series

Volume 1   Technologies and markets
           J.J. Verschuur

THE BUSINESS OF ELECTRONIC PRODUCT DEVELOPMENT

Fabian Monds

Peter Peregrinus Ltd on behalf of the Institution of Electrical Engineers

Published by: Peter Peregrinus Ltd., London, UK.

© 1984: Peter Peregrinus Ltd.

British Library Cataloguing in Publication Data

Monds, Fabian
    The business of electronic product development.
    — (Management of technology; 2)
    1. Electronic industries — Management
    I. Title   II. Series
    388.4'7621381'068   HD9696.A2

ISBN 0-86341-012-X

Printed in England by Short Run Press Ltd., Exeter

Acknowlegments

Photographs are reproduced by kind permission of the following:

The Institution of Electrical Engineers (4.5., 4.6., 4.7)
Sinar Agritec Ltd. (5.3)
Hekimian Laboratories Inc. (7.2)
Apple Computer (UK) Ltd. and Mr Robin Jelly,
Medical & Scientific Computer Services Ltd. (7.5)
Standard Telephones & Cables plc. (8.3., 8.4., 8.5 and 8.6)
British Telecom (8.7)

Contents

		Page
Preface		**x**
Acknowledgements		**v**
1	**Introduction**	1
2	**The dimensions of the business**	**4**
	Dimensions and disciplines	7
	The engineering dimension	8
	Management	10
	Marketing	12
	Money	14
	Some evidence	15
	Encountering the business dimensions	17
	Moral	18
	Product development work	19
	References	19
3	**Product development process**	**21**
	The project and the product life cycle	23
	The nature of the project	25
	Development people	28
	The entrepreneur	29
	Project manager	29
	Design engineer	29
	Production engineer	30
	Technological gatekeeper	30
	Marketeer	30
	Controller of resources	30
	References	30
4	**Project phases**	**32**
	Concept and definition	32
	Feasibility study and the business proposal	36
	Design and engineering	39
	Technological environment	39

Design factors   45
Product specification   45
Overall product objectives   46
Team and tasks   46
Practical procedures   47
Engineering aspects   49
Decision time   50
Transition to production   50
Comment on project phases   51
References   52

5   **Feasibility study**   53
Feasibility   53
Market assessment   58
Technical assessment   60
Technical feasibility study   65
A feasibility example – Sinar Agritec Ltd.   66
Corporate interaction   68
Decision or proposal?   72
References   73

6   **The business proposal**   74
New product development programme   77
Microwave security system VMS   77
The financial plan   79
Help from the computer   80
Historic accounts   80
Development cost   81
Projected profit and loss accounts   81
Break-even chart   83
Projected cash flows   85
Balance sheet projections   90
Assumptions   91
The proposal's merit   91
Sources of finance   92
Self-help   93
Bank overdraft   94
Bank loan   94
Government grants and loans   95
Venture capital   95
Financial packages   96
The presentation   96
References   99

7   **Project management**   100
The need to plan   100
Project management principles   102
Objectives   104
Hekimian Laboratories Inc. Case Study   106
Background   106
Extract from specification for Model 3934   106
Development sequence   107
Comment   108

Planning and control   108
Guidelines for small teams   111
   Specification   111
   Tasks, signing up and estimates   111
   Plan   112
   Control   113
   Personnel   114
A product development parable   116
Techniques   117
Millimeter Wave Technology Inc. Example   120
Organisation   121
Testing and transition   122
Self-defence   124
The spirit of product development   125
References   126

8   Ear to ear and hand to mouth: the story of a new electronic telephone   127
Concept and definition   127
   Principal design criteria   128
   Product development strategy   128
The definition contract   129
The technical development project   130
   Mechanical   132
   Electronic   134
Transition to production   134
Final comments   136
References   138

Index   139

Preface

The creative business of electronic product development is stimulating and rewarding. The process stretches from concept to commercialisation, and participants of many disciplines share the sense of achievement when the new product is successfully launched on to the market. This success depends on good engineering and good business management. It is commonplace to contrast technical and commercial approaches, but for fruitful electronics innovation, engineering and business perceptions are complementary and mutually supporting. The bipartisan approach, which is the foundation for constructive and effective interaction of engineering and management, is a principal theme of this book.

The objective is to provide information and guidance for those working in electronic product development, particularly the project manager and professional project team members. Such individuals may have engineering or managerial qualifications and background; it is hoped that the engineering colouring of management topics is appropriate to both interests. The engineer taking on project management responsibility encounters unfamiliar disciplines and many challenges; this book addresses the common ground of electronic product development, where engineering, management, marketing and financial professionals plan and work together. Innovation, technology and management combine in the definition, design, development, marketing and manufacture of the new electronic product. Every professional engaged in this endeavour has a managerial responsibility, if only for his or her own work. Project teams are usually small with a limited number of key contributors, and effective management is often very dependent on the team member's contributions to planning, scheduling and control. The needs of the project team manager and the team members are considered, using the product development sequence as a chronological basis for the presentation and discussion of topics. In the ever-changing world of electronics, the entrepreneur has a special place. Attention is given to the strengths and needs of the new small company built upon innovation in electronics.

In this book, the several dimensions of the business of electronic product development are introduced, the development process is described, and the major phases of the process are delineated. The theme of engineering project management is elaborated upon to examine in detail feasibility study and the business proposal, and the design and engineering work which constitutes the technical development project. Management methods for the development programme are discussed and advice is offered to managers and project team members. Frequent reference is made to illustrative examples of achievement and difficulty.

The presented material is based on study, dialogue, observation and personal experience.

Fabian Monds
Enniskillen, County Fermanagh, September 1983

Chapter 1

Introduction

There are people who can do things and there are people who can get things done. To be more specific and a little less homely, engineers take pride in practical accomplishment and the solution of technical problems, and managers value effective planning, organisation and control of work to achieve objectives. The business of electronic product development needs people with both sets of skills and values.

New product development involves many activities of a specialist nature; in the electronics industry high standards of precision and reliability are demanded for design and engineering, which often claim star billing when success stories are told. Nevertheless, product development is carried out for a commercial purpose. The business of product development is decisive for every electronics company, since it is likely to be the principal determinant of prosperity and survival. Commercial and technological factors combine to influence the outcome of a new product programme – their interaction is more predictable if both are considered right from the product's conceptual beginnings. Business judgment should never be far from technological endeavour.

The business aspects of product development include overall commercial strategy for the product, involving finance, marketing, production, sales and service, and also detailed financial management of design and engineering project work. The sound business principles of delivery on time, to specification and within budget, are as applicable to product development as to mass production. In the electronics industry, the rapid pace of technological change and the dynamism of the marketplace give enhanced significance to business activities other than manufacturing. The activities of product planning, development and marketing, with the commercial feedback that links them, are seen clearly as critical investment areas. The return on this investment is rather less certain than the well defined benefits provided by modern production and test facilities, whose worth is unquestioned.

Indeed, commitment to new product development represents speculation. The probability of commercial success is heavily influenced by the quality, availability and utilisation of human, physical and financial resources. The very newness of a new product means that those developing it have to be prepared to do at least some things differently and to learn quickly from their mistakes. Obviously, development programme planning and control is not particularly amenable to formal methods, but organised efforts to achieve objectives are still needed. Technical and managerial skills must be applied in harmony to optimise the effectiveness of resource deployment and utilisation in the cause of product innovation. This is true for every industry, since every industry must innovate, but innovation − 'the introduction of something new into the established order' − is a definitive characteristic of electronics. This industry and its new products need customised management techniques as well as custom chips.

As a general industrial topic, new product development is conventionally viewed from two standpoints − the technological and the commercial. Often the resulting perceptions are contrasting rather than complementary and reflect the preconceptions and even prejudices of the observers. The challenge, excitement and intellectual demands of technical development work can lead the engineer to question the relevance and value of business management, especially when the interaction of management with engineering is biased towards financial monitoring and control. Negative impressions can also dominate managers' attitudes to engineers, who may come to be regarded as practitioners of obscure but essential black arts, prone to unpredictable behaviour. There may be some truth in these opinions. Engineers may know how to handle 'things', but can have poor judgment of people and priorities. On the other hand, managers' appreciation of technology's opportunities and limitations may be inadequate. The companionable epithets of 'techie' and 'fuzzie' are suggestive of the gulf that can exist between engineer and manager.

Professional narrow-mindedness has been the cause of limited achievement in many a business. Fortunately, the always youthful industry of electronic engineering has suffered less from such conflict than other, older industries. In many companies, electronics engineers have taken on top management responsibilities and business managers have responded with enthusiasm and commitment to the new opportunities and challenges of electronic technology. These are demanding undertakings, since electronic engineering, always in transition, has many new techniques and tools at its disposal, including the power of computer-aided design, development, manufacturing, test and management. Innovation in electronics requires that both engineering and business aspects are given appropriate attention: managers must understand the implications of commercial decisions for technological progress; engineers must appreciate the business dimensions which delineate their technical environment. These positive attitudes can prevail when managers and engineers are prepared to communicate freely and constructively and see the other's point of view. There are sometimes limits on the degree of insight which managers can have for design and engineering topics, without specialist education in electronics, but awareness of the patterns and pressures of technical product development work contributes to effective management. Of course many managers have excellent technical backgrounds, and these qualified individuals are able to maintain the technical−commercial balance. Best of all, engineers can become managers, without abandoning engineering, by dint of work, experience and opportunity. The first steps along this road can be taken in the context of product development, in the management of a personal work programme or the interaction of task groups.

The way engineers work and the ways they manage and are managed are changing. More direct contact with customers, marketing people and financial control is one aspect of the trend, another is the freedom for self-management provided by new computer-aided design and engineering tools. These considerations demand that engineers are business conscious, alert to cost considerations and perceptive to market needs and that their work is organised to satisfy not only the technical product requirements but also the commercial objectives.

The ideal product developer needs a further talent to add to engineering and management. It is a measure of entrepreneurial ability, the quality that 'makes new business happen'. The commercial components of new product development need not be constraints or impositions; they can be part of the innovative structure, amenable to optimisation and exploitation. The rare combination of technical—managerial—entrepreneurial product innovator does exist and has business achievements to prove it, but even if the engineer or manager decides to stay on his or her side of the business fence, awareness and appreciation of the range of skills and resources necessary for successful product development will help communication and co-operation.

The technical and managerial tools and techniques of product innovation can be applied in many environments, from the entrepreneurial business, where everything depends on a product success, to the development unit of a large company or an academic institution, where the financial consequences of success or failure are not immediately evident. Despite the variety of situations, it is very probable that development work on a product is carried out by a small group. Organisational structures and procedures for such groups are therefore of wide applicability. A 'small business' attitude to electronic product innovation can be effective in even a giant corporation.

In the chapters which follow, these thoughts are expanded and illustrated. There is also some advice on how to handle the several dimensions of the business of electronic product development.

Chapter 2

The dimensions of the business

Electronic product development has both technological and economic dimensions. It is sometimes tempting to consider only technical aspects – engineers are vulnerable to this temptation. Elegance of circuit design, cleverness of component choice and artistry of printed circuit board layout are but a few of the engineering features which can fully occupy an engineer's attention. The cost effectiveness of the product design is often of rather less interest, but economic parameters dictate the very environment within which product development takes place. Indeed, the success of today's projects will greatly determine the scale or actual existence of tomorrow's new product development programme. Commercial considerations must feature in new product development planning and execution. The business of electronic product development has many of the commercial features of any product development based on modern technology. Lessons learned in other industrial spheres can be applied to the task of managing innovation in electronics. However, the treatment of product development varies widely from industry to industry. The integration of R & D (to use the most common descriptive term) into a company's total business is one of the most distinctive features of the electronics sector. Another good example of R & D integration with business is in the aerospace industry. Specialised management techniques, notably for project planning and control, have been developed in these industries to cope with the dynamics of the business. For aerospace projects, with literally thousands of activities, planning has to be sophisticated and very extensive. Electronics projects can use a wide range of planning methods, many quite basic and easy to operate. Furthermore, the electronics industry is unique in that the same technology which is built into our new products is used so extensively in the equipment which can assist in their design, manufacture, testing and maintenance. Micro, mini and mainframe computers and information technology aid project work and the management of that work. The designer of electronic products is best equipped to exploit his 'native' technology in the product and for the product. Hence rapid progress in utilising new equipment and new procedures is the norm.

In contrast to other industries, electronic product development programmes can be quite short term; one or two years can span major projects. Electronic product development has, in relation to most other industries, low total project costs and short project durations. Small teams usually carry out entrepreneurial electronic product development. One or two individuals can be responsible for highly innovative and commercially successful products. Just one example is the Apple personal computer created by Wozniak and Jobs (see References 1 and 2). Stephen Wozniak designed the Apple I microcomputer in his spare time while employed at Hewlett-Packard. His friend Steven Jobs convinced him that the product could be a commercial success. In 1976 they invested $1300 to build the first machines with Jobs looking after manufacturing and marketing while Wozniak continued development work which produced the Apple II. They were joined by A.C. Markkula, the former marketing manager at Intel Corporation, who offered his expertise and a personal investment of $250 000 in the project. With a good product and an effective management team, the company grew very rapidly, with sales of $200 million in 1980, $335 million in 1981 and $583 million in 1982. Apple went public late in 1980, and major investments have been made in new product development, notably the LISA launched in 1983.

A UK personal computer business success, in this case with a minimum price product for the home, has been that of Clive Sinclair. The ZX81 home computer achieved monthly production and sales of over 50 000 units within a few months of its launch. The innovative design lent itself to economic production, but the ZX81's success has been attributed primarily to marketing brilliance [3]. The Sinclair phenomenon has been defined as the decision 'to enter those races that are worth winning but that no one else even knew were going on'. While the definition may lack some of the logic that has been so effectively built into the Sinclair product, the emphasis on marketing and business acumen is unmistakable. Sinclair Research Ltd. retains small business virtues and achieved profit earnings of £14 million in 1982 with a staff of only 60 direct employees. Manufacturing of the product range has been subcontracted, involving the support of about 2000 further jobs.

These success stories illustrate the multidimensional nature of the electronics business. A categorisation of the activity involved in electronic product development into technical and commercial dimensions will permit the discussion of principles and procedures in an orderly way. Of course all the dimensions are significant all of the time; also they interfere with each other and are affected by external events. Later these interactions can be considered, by reviewing some actual development experiences. So bearing in mind the existence of interaction, integration and co-ordination, consider what the dimensions are.

Obviously engineering is the first dimension for the professional engineer. It encompasses a wide range of operations — for example, technical feasibility study, manual and computer-aided circuit and system design, prototype fabrication, production and testing, service and documentation, but technical achievement is only one of the requirements for product success. Many studies of general technological innovation have attempted to assess the likelihood of commercial success for new products and to identify the critical factors involved [4, 5]. Given the wide range of industries examined and the variety of starting points for assessment (when does a product development project actually start?) generalisation is dangerous. However, quoted success rates are usually in the 1% to 5% range, so it is clear at least that the odds are heavily against happy endings. This is not so surprising since the outcome is determined by the joint probability of technical design completion, commercialisation *and* market success.

If the probability of technical design completion is 0·5, the probability of commercialisation given design completion is 0·5, and probability of market success given commercialisation is also 0·5, the overall success rate is 12·5%. In practice it is unlikely that probabilities as high as these are achievable over a sample set of projects. By commercialisation what is meant is the transfer into production and the launch onto the market of the new product. Management decision-making obviously affects the probability of this taking place.

As regards critical factors, limitations on technological capability are rated relatively low in significance — in so far as they represent constraints on successful new product development. More significant factors are frequently reported to be organisation, marketing and finance. These three, conveniently identified as management, marketing and money, are dimensions as important as engineering in the business of new product development.

The four dimensions:

> Engineering
> Management
> Marketing
> Money

constitute not only product development but also the totality of the engineering business.

Each dimension is indispensable — and each one means work. It would be quite possible to discuss at length the constitution of these dimensions, in terms of component proportions of art and science, practice and profession. A simpler approach is to concentrate on the objectives and activity involved in each and on the commitment required from their practitioners. It is not meaningful to try to rank the dimensions in importance — it is true that management is what holds every institution together and makes it work, as Peter Drucker puts it [6], but the work involved in coping with any of the other dimensions can be just as demanding. The work is carried out by people, and each of the dimensions requires effective contribution of effort by competent and committed professionals.

Of course the management of the total business encompasses strategic planning and day-to-day control, while total engineering activity includes manufacturing and quality assurance as well as product development. Product development's dimensions differ in their constitution from those of the complete business, in terms of the relative importance of specific components, for example manufacturing considerations. Nevertheless, the component activities of the total business are all important to product development. Their characteristics and relative significance for the individual staff member depend on his or her job responsibilities. At both levels the dimensions are not separable. Cost considerations influence manufacturing and marketing programmes as well as development project planning; marketing strategies affect production and quality assurance functions as well as new product specifications. Management must co-ordinate and integrate all the component activities, allocate and control resources, so that the business survives and prospers.

Dimensions and Disciplines

The business dimensions can be thought of as disciplines, although it is obvious that professional specialisations are generally rather more precise — engineering encompasses production and quality control as well as R & D, to name but a few. A problem area in the engineering industry can be communication between disciplines and specialisations, or rather between their practitioners. A successful new product requires achievement in all dimensions yet professional narrow-mindedness can jeopardise such achievement. For example, the communication gap between engineer and marketing specialist, amounting almost to a clash of cultures, can be a major obstacle to progress. Communication must be made effective, bridging such gaps, however tempting an insular attitude may be. Appreciation and understanding of all dimensions of the total business is essential for the top management of any company or concern; it is just as important that those responsible for the business of new product development within a company or concern can cope with the multidimensional nature of their task. To become more specific, the dimensions are listed in Table 2.1, with representative functions at both total business and product development levels.

In new product development, engineering and marketing are the sources of technical and commercial creativity and innovation. Management and money are the integrating, co-ordinating and facilitating dimensions. In this book, management and money aspects are given more detailed treatment than engineering design and general marketing, but the importance of all four dimensions is unquestioned. Technical aspects of electronic design are considered only generally and in the context of product development example situations. There are many excellent books on every electronic design specialisation and the professional engineer is well aware of the need to keep up to date with the technology.

Marketing is an important part of product development — not only as a driving force and a mechanism for assessment of product suitability, but also as a weapon in the engineer's armoury. New products need support within the company or organisation, to ensure that deserving projects are approved, that development contracts are placed and that new products are launched with the full backing of top management. The 'product champion', be he from the ranks of engineering, marketing or general management, uses marketing techniques to achieve acceptance and success for his product.

The published literature and recorded knowledge on each of the business dimensions is extensive. The approach adopted here is first to provide an outline of the

Table 2.1   *Components of the business dimensions at corporate and product development levels*

Level	Engineering	Management	Marketing	Money

Corporate or total business	Facilities
	Manufacturing
	Quality assurance
	Service and
	  maintenance
	Training
		Planning
		Organising
		Staffing
		Directing
		Controlling
			Product strategy
			  and planning
			Sales pro-
			  grammes
			Advertising
			Market
			  surveys
				Ensuring resources
				Management accounts
				Financial control
				Financial modelling
				  and analysis
				Corporate planning

New product development	Specification
	Feasibility study
	Design and
	  product
	  engineering
	Documentation
	  for production
	  and test
	CADMAT
		Project
		  manage-
		  ment
		Planning and
		  scheduling
		Staffing
		Task assign-
		  ment
		Motivating
		Leading
		Crisis control
		Paperwork
		Liaison
			Product concept
			  and definition
			Market research
			Feasibility study
			Product launch
			Product promo-
			  tion
			Derivatives
			Reviews
				Development budget
				Contract negotiations
				Project planning and
				  financial control
				
				Project auditing
				Funding of new
				  capital equipment
				Grants and loans
				Contingency planning

dimensions' significance in the general business context. The structure for product development is the project. Accordingly the managerial, marketing and financial dimensions will then be examined principally in the context of electronic development project management.

The engineering dimension

Engineering has to do with the application of science and technology — more basically, it is concerned with making things. In this book there is a concentration on electronic engineering, but even in the electronics industry it is very unlikely that this specific discipline will be practised in isolation, with no interaction from other engineering sectors. Electronic products need packaging for sale and use and consequently most of them incorporate mechanical components and structures. Materials technology helps with the engineering of switches, keyboards, covers and connectors. Robotics and automation contribute to efficient production. Software is needed for microprocessor and computer based products. The electronic engineer has to pay attention to the interfaces between his design skills and other engineering disciplines, as well as to the interaction of the management of his project work with the operation of the complete business or organisation.

Within the product development project, design work is, of course, pre-eminent. Electronic circuit and system design cannot be carried out with a simplistic emphasis on mere functionality – the designer must take into account the required product performance in the market-place. This includes product appearance and ease of use – aesthetics and ergonomics. Reliability of operation in the anticipated environmental conditions, maintenance needs and safety considerations have to be addressed. The commercial performance of a product may well depend on the ease and economy of its manufacture – is it to be assembled manually or on an automated production line? What quality control, test and detected fault repair procedures are optimum? What design standards and codes are to be adhered to? Documentation must be comprehensive and usable. Cost consciousness must pervade the design activity. Can supply of components and services for production be assumed? The designer or product engineer cannot relax when the transition to production has been implemented. Experience with the product in manufacture or in the market place may generate the need for engineering changes, often at very short notice.

These basic requirements for product design are well known, but today the pressure is on for increased design productivity, with maintenance of quality and cost effectiveness. Product life times have shortened, notably in consumer electronics, while product complexity and sophistication have increased. Development time also has to be reduced, to match the dynamism of the market place. Even when a product has an anticipated market lifetime of several years, the threat from competitors may forcibly compress permissable development timescales.

These considerations, together with the increasing importance and availability of complex very large scale integrated (VLSI) circuits, as standard, custom and semi-custom chips, have led to the adoption of computer-aided techniques. The acronym CADMAT – computer-aided design, manufacture and test – has been adopted by the industry in the UK to indicate the total applicability of the approach. Using CADMAT, product designs can be developed and their operation simulated. Conceptual circuits can be converted into designs using available components and devices. Printed circuit board (PCB) layout can be performed almost entirely automatically and data for direct use in manufacturing, including information for PCB drilling and automated component insertion, can be generated. Test programs and documentation can be produced immediately.

All this and more is possible with a CADMAT system, often under the direction and control of a single engineer at his workstation.

The design engineer must be able to organise his or her work. This includes management of the selection and use of technical resources. Normally a group of engineers will work together on a product development, with tasks assigned in categories such as analogue, digital, packaging; hardware, firmware, software; quality and documentation. Such engineering groups need management to provide planning, control and leadership.

The engineering dimension is thus correlated with the other dimensions of business.

Management

The management of electronic product development requires the use of techniques which are generally applicable, but the special features of the business mean that these general techniques are not the complete answer. The dynamism of electronics shows no sign of slackening and management must keep pace. Often the roles of design engineer and manager are inseparable. The task of managing the business of electronic product development is a shared responsibility — shared between engineer and business manager, designer and project leader, technical, financial and marketing specialists. Specialist technical resources will be deployed. The special nature of electronic projects must be considered. Sound management principles should be applied.

The functions and activities of management have been defined by many authorities [6, 7]. In giving a version of them here, it should be emphasised that the management of product development requires not only their application, but their application in a very dynamic situation. Reappraisal of plans and close monitoring and control are essential for success. In effect, sound management principles must be complemented by flexibility and continuous attention to progress and to the significance of new information and results [8].

First, the functions of management are as follows:

1 *Planning*   Objectives must be defined and the work to be done determined. Plans must be made whereby the objectives are to be achieved.
2 *Organising*   The work has to be divided into activities and tasks for effective utilisation of human and technical resources.
3 *Staffing*   Personnel requirements must be defined and suitable people employed to carry out the work in accordance with plans and organisational structures.
4 *Directing*   Human and technical resources have to be guided and commanded to achieve objectives, using appropriate information systems for communication and co-ordination. Directing involves motivating and leading staff.
5 *Controlling*   Performance must be monitored and assessed and performance improvements must be obtained where necessary, so that objectives are achieved.

This set of management functions begins with planning and ends with controlling, but, as indicated by Fig. 2.1, planning and controlling are linked in the basic 'management loop'. Activities must be controlled for plans to be carried out so that objectives, the desired results, are achieved. In every business plans must be modified in the light of events — external influences or internal changes. This is particularly so in product development where a new discovery, a new procedure or a new component can necessitate changes in plans. It may even be desirable to change not only the plan for achievement of objectives, but the very objectives themselves. For example, the availability of a new microprocessor may allow additional functions or performance features to be incorporated into a new product, or new information from the market-place — perhaps about the activity of a competitor — may cause a revision of original objectives.

The definition of proper objectives and the adoption of methods which ensure their achievement – even though changes may become necessary in the desired results and corresponding plans – are key business processes.

This approach to management is sometimes referred to as 'management by objectives' (MBO) or management by objectives and results (MOR) to emphasise that control to achieve objectives is as important as the objectives themselves [9, 10]. A plan without implementation is of little value.

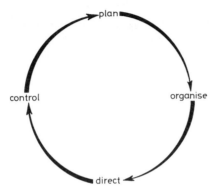

Fig. 2.1.   *The management loop*

Management by objectives and results is an effective approach to the business of product development. MOR principles can be adopted for development project planning and control. Of course the other three management functions – organising, staffing and directing – must also be performed properly. All five functions are the responsibility of the project manager or project leader, who must perform them with an appreciation of the need for co-ordination and integration.

Management needs resources – human, physical and financial – to carry out its functions. In electronic product development these resources are often special; they differ from those required for production and maintenance. The money stays the same, but its allocation to product development and the control of its use there needs special attention. Management planning and control functions must give financial aspects, particularly budgeting, as much emphasis as work scheduling and technical achievement.

The application of planning and control techniques to project management will be considered in more detail later. An essential prerequisite for effective product development is indeed the definition of proper objectives. What is to be achieved and when? A useful approach to objective definition is to be clear as to who the 'customer' is, and to have the customer closely involved in planning at the earliest possible stage. Close attention to the customer's requirements can help prevent excursions into technological mazes or blind alleys. In this respect, management needs marketing.

A large scale illustration of the cost of inadequate attention to the customer's real needs was the Bell System 'Picturephone' project [11]. Existing telephone lines were used to convey monochrome limited definition video signals as well as audio, so that callers could see as well as hear each other. Over $100 million was reportedly spent by the company, notably on work at the Bell Holmdel Laboratories in New Jersey. The system was launched in the mid-1960s, but failed to gain market acceptance – apparently for a number of reasons, including human shyness! Bell withdrew Picturephone as a domestic service. Today, emphasis is being placed on video-conferencing, for business and government, where market needs can be well defined.

The functions of management and the importance of management by objectives have been indicated, but methods work only if people apply them, and management has been defined as 'the art of getting things done through others' [12]. Successful product development depends on people, perhaps to a higher degree than in other industrial activity. Creativity and innovation are needed, but cannot be planned. Management planning and control are not enough – motivation and leadership must be given to the individuals in the development team and the right environment established.

Management must emphasise human qualities and values. In particular, communication is needed, to convey information on objectives and plans and to ensure that necessary modifications to those objectives and plans are made in time. Management motivation and leadership can provide the sense of purpose which drives the development team forward to success. Effective communication of this sense of purpose requires an approach which has many of the features of marketing, in the most general sense.

Marketing

The development of the Data General Eagle 32-bit computer, later rechristened the Eclipse MV/8000, is dramatically recounted in Tracy Kidder's book 'The Soul of a New Machine' [13]. This US best seller was the 1982 Pulitzer prize-winner – perhaps surprisingly for a book about computer engineers and their work, containing a considerable amount of technical explanation. Maybe electronics and computer technology now has a good 'market image'. There is no doubt that the growth in sales of electronic consumer goods, including personal computers, has been accompanied by heightened public interest in the technology. 'The Soul of a New Machine' describes the human relationships, pressures and achievements, including the project leader's continual attention to marketing concepts, to ensure the product's eventual commercial success. He also had to 'sell' the development project within the company and this required much effort. Competition for resources including funding was intense – the Eagle group had to convince top management that their proposals and work merited continued support: 'Let's show 'em what we can do.'. This commitment to commercial success and project promotion is a good illustration of the truth that technical achievement needs the addition of management, marketing and money if a new product is to succeed. Data General's project leader Tom West is described as engaging in promotion within his group to 'stir up enthusiasm' and as selling the new computer to other parts of the company, first as insurance against failure of other development projects and later as a 'flagship' product, 'He was proselytizing for Eagle throughout the company.'.

The Eagle example shows the importance of marketing – marketing which relates to the product in the market place and marketing which is realised by promotional effort within the company or organisation. In both situations customers must be convinced. Marketing has been defined as 'knowing the customer, identifying, anticipating and satisfying needs and desires, researching and analysing customer behaviour [14]. Indeed its function has been described as creating the customer.

At the 'total business' level, marketing contributes to effective product planning and strategy – what products should the company be developing, manufacturing and selling? [15]. New product opportunities are sought out and screened, with product development undertaken for probable 'winners'. There is a wealth of literature on product strategy, mainly from a marketing viewpoint. It is generally agreed that the key sources of new product ideas are research and development and market search. This is particularly so in the electronics industry, where technology 'push' and market 'pull' have been harnessed together to great effect. Just one of the multitude of successful development and marketing collaborations has been Texas Instrument's Speak and Spell product – the hand-held speech-synthesiser based learn-to-spell aid. As reported [16] the initiative for this product came from four engineers, who took their idea for a new electronic learning aid to the company's Objectives, Strategies and Tactics Committee in late 1976. The project was granted $25 000 for three months to assess feasibility, which was demonstrated by a computer simulation of the proposed 1200 bit per second speech synthesiser integrated circuit. 'Parallel to the efforts of selling solid-state synthetic speech to management was laying the groundwork for marketing the new talking learning aid'. Again, selling externally and internally was required. Speak and Spell, launched in June 1978, has been a very successful product.

Technical and marketing information is used in the product search and identification process [17]. Product definition is assisted if the product function is emphasised – what must it do for the customer? When a product has been defined a development programme is needed. Marketing must feature in the planning and control of the project. Marketing attitudes must influence the conduct of internal communications – within the development group and throughout the company or organisation – so that appropriate support is available. The chances of the product being right for the market-place, and of it gaining acceptance and enthusiasm from the production team, the sales force and top management, are thus maximised.

It is inappropriate to dwell on marketing as a specialist discipline in a book of this kind but the importance of marketing considerations, information and attitudes in the conduct of electronic product development programmes will be seen to be pervasive.

Money

Every company, organisation, institution or concern needs money to operate. The money can come in the form of payment for products or services provided or as investments, loans, grants or gifts. Product development activity requires funding, but unlike production or marketing, the return on money invested in it is not usually visible for some time. At the end of a successful product development programme, it is possible to be very specific and quantitative about financial requirements for production and selling operations. At the beginning, financial planning is much more difficult. During the project, as information is gathered, plans may have to be changed. Increased financial requirements are not popular. Accordingly, the entrepreneur or product development project manager must be very careful to generate adequate financial estimates and budgets and to operate within predicted limits of expenditure. Financial planning and control are essential. Without them, product development projects will not even be started.

Although the product development group does not have products to sell, it can and must sell its capabilities and ideas. The 'customer' must be persuaded to provide funds so that work can be done. Again, management and marketing dimensions are closely linked to engineering and finance. The product development process must commence with a proposal, which has to include financial considerations to be sensible and assessable. It is clearly best if commercial marketing considerations feature in even initial proposals.

If the product idea has originated in engineering, which happens frequently in electronics, marketing aspects should be carefully examined. Products which represent a solution in search of a problem rarely succeed – there must be a customer need, even if the need has to be created.

There is a wide range of situations in which a product development business proposal can be generated. The individual entrepreneur may wish to find a financial backer, the group manager may want to initiate work using newly available devices or the corporate planning of a company may have determined that a product to a particular functional specification is required. The financial appearance of the resulting proposals varies accordingly. However, each one has to address the same 'bottom line' – cash requirements. The form of presentation has to be understandable by both engineering and financial staff. Although financial analysis can be very sophisticated, basic techniques can provide a great deal of useful information, usually sufficient for decision making and monitoring. The business proposal can be based on a cash flow summary, showing cash movements period by period and identifying expenditures, accompanied by predictions of profitability and balance sheet projections. The manager of a product development group in a substantial company or organisation will receive help in financial management from accountants and other specialists. Nevertheless, a commercial attitude should be adopted in planning and control activities, to ensure that financial resources adequate to the task are available, and that operations are within agreed budget limits. Group financial reports and records must be kept up to date – in large organisations delays through the financial reporting system may mean that budget overruns are discovered too late. The financial information needed by the project manager is likely to be generated within the project in the form of engineers' time sheets and records of equipment, supplies and services ordered.

The small business – perhaps formed by two or three individuals taking their first steps into the business world – requires comprehensive business planning. In the early days, the very existence of the company may depend on one successful product development. A cash flow summary, however detailed, which is limited to the expenditure involved is inadequate to support the raising of finance or to provide for management of the business. The cash flow must be extended to include income from sales and grants. The needs of the ongoing business, notably working capital to cover product launch, production set up, stocks and the delay between purchasing expenditure and sales receipts, must be quantified. To satisfy a financier, the business plan will usually have to cover 2 to 3 years and include projected profit and loss operating statements and balance sheets, even though prediction may be very difficult. The engineering entrepreneur can cope readily with such requirements – he can learn about the limited accounting procedures involved and implement them. The electronics engineer is particularly privileged, with his familiar companion the personal computer to assist in bookkeeping and management accounting. The engineer's sense of responsibility for achievement to time and quality standards has to extend to financial matters. Money matters relevant to electronic product development are discussed in Chapter 6, 'The business proposal'.

Some evidence

The definition and significance of the business dimensions: engineering; management; marketing and money, may appear to be obvious and based on common sense. However, acceptance of principles does not ensure their application. There is considerable evidence to support the belief that companies and organisations which carry out new product development with real attention paid to the four dimensions are likely to succeed. There have been many studies of success and failure in product innovation. Some of the findings and factors from general studies which are relevant are as follows:

'It is still not recognised by enough people in industry just how essential good design is to the success of the business' [18].

'As the development of new products necessarily involves most functions and departments in a business, communication and co-ordination are critical' [4].

'The great part played (in achieving success) by personal leadership and peoples' interrelationships' [5].

'Businessmen (who) perceive a human need and seek to serve it profitably by mobilising management, capital and the works of science' [19].

'It is a vital management task to find the right people and to set up structures that will enable them to co-ordinate all the activities necessary to bring a new product successfully to market' [4].

'The dire results of an inadequate strategy brief and organisational confusion' [4].

'Integrated project planning from idea inception through development to marketing' coupled with 'Planning methods simple and flexible enough to deal with the incentives of new products' [17].

'The crucial necessity of establishing that there is an actual or potential market demand for the product' [4].

'Successful innovators have a much better understanding of market needs' [20].

'A formal management system is not sufficient in itself but must be associated with the efforts of individuals highly motivated for success – 'technological entrepreneur' or 'project champion'' [8].

'The three major areas of constraint (in new product development) are finance, markets and internal company matters' [4].

'The clear and unambiguous allocation of accountability' [21].

'There are three things that must be done . . . check out the produceability by your company . . . and do the necessary market research. The third thing is run the numbers' [22].

'Successful innovations are allotted sufficient cash and manpower resources . . . at critical stages in the process, successful innovators focus resources onto the innovation to facilitate its progress' [23].

Many such statements and points have been recorded, supporting the importance of engineering, management, marketing and money in new product development. Several detailed analyses of key factors have been produced, indicating critical points of general applicability to product development. Work which has paid special attention to the electronics industry or has separated out results for electronics is of particular interest here.

The impact of the microprocessor and government appreciation of its significance has encouraged considerable study in the UK of skills and methods relevant to product and process development based on the technology [24–26]. In the 1980 study funded by the UK Department of Industry and carried out by the Computing Services Association, 120 companies using microelectronics and microprocessors in products, processes and information systems were surveyed. The study team found that in the great majority of projects, two factors were major contributors to success:

clear identification of objectives;
management commitment.

In addition, the involvement of relevant departments, good communications and input from marketing were found to be important.

Professor Ernest Braun, in an earlier survey paper of the US microelectronics industry [27] states that 'the most successful people in the industry are those who combine commercial acumen with sound scientific and engineering know-how'.

Some of the most thorough and well organised studies of new product development have been carried out by Professor William Souder of the University of Pittsburgh. In a study [28] which examined 50 successful and 50 failure outcome projects in eight industries, including electronics, it was concluded that 'the degree of harmony, joint involvement and felt partnership between R & D and marketing were significant determinants of project success or failure'. The most successful innovating firms used 'teams, task forces and project systems to manage the R & D/marketing interface'. On the eight industries, electronics ranked sixth in project cost and eighth in project duration (average 1.58 years), that is, electronic product development is fast and inexpensive compared to other industries (for example, metals, transportation and machinery). However, the average cost of failure outcome projects in electronics was approximately three times the cost of success, because of 'expensive and unsuccessful attempts to correct defects in the products at the production stage, because of attempts to accelerate the completion dates of projects which were behind schedule, or because of an inability to terminate unsuccessful efforts in a timely fashion'.

An even more extensive review of the electronics industry would probably show up a very wide range of product development project cost. Development of a new personal computer will require many times the investment needed for a hand-held electronic toy. Apart from such a qualification on total project costs, Souder's work, like that of others, clearly emphasises the significance of management, marketing and finance as well as engineering in the product development process.

Encountering the business dimensions

A small company was founded by a group of electronics engineers and computer scientists to develop, manufacture and sell interface systems with customised software. The products were sold to customers in scientific and industrial laboratories who wanted to link instrumentation to mini and mainframe computers. After about a year the company had shown considerable potential for success and a financial loan package for expansion was negotiated with a bank and a government agency. One of the conditions for support was that a firm of management consultants would be engaged to set up a bookkeeping and financial control system. This turned out to be quite expensive for the company, but, more seriously, required the involvement of every member of the firm's staff (five at this stage).

The management consultants knew a great deal about bookkeeping and financial control, but very little about interfaces and software, and a lot of time was spent by the company's staff, particularly the owner directors, explaining the specialist development, manufacturing and selling activities. After many weeks a comprehensive financial control system was defined, documented and put into operation. At the same time it became clear that over this period delivery deadlines had been missed, development problems were not being solved and new orders were non-existent. In something of a panic, the directors had a meeting with the management consultants. One director's memory of that meeting is that the most significant statement (made by one of the financial specialists) was 'if the marketing is not right, nothing is right'.

This company had started off with one out of four right – that is, the technology. Management, marketing and money did not need great attention when the firm was dealing with only a handful of sympathetic customers who paid promptly. With expansion came the need for a more professional approach to the business and this was very properly insisted upon by the loan providers, but the choice of mechanism to establish better financial control was poor. The management consultants had little experience of innovative technology based business and time was wasted educating them. The firm's directors did not deploy their resources efficiently to ensure that business activity was maintained at an effective level while the new control systems were being devised and set up. One director should have been selected to liaise with the consultants; as it happened, everyone got involved. With the establishment of the financial control system, the company had two out of four right. The meeting with the consultants reminded the directors of the importance of marketing, and shortcomings in that area were put right by the efforts of one director who took on the task; three out of four right.

The decision to expand the company and the assistance sought and obtained had nearly led to the collapse of the business. The 'near miss' caused the owner directors to review their objectives, plans and control procedures – in other words their management approach. All four dimensions of the business were then being given proper attention.

After this trauma, the company prospered and expanded to more than 30 staff, with new products introduced as electronic and computer technology and markets changed. The financial control system was modified to suit the changing business, and transferred on to a small computer.

Moral

Founders, owners, directors of new technology based firms – the entrepreneurs – are responsible for *all* of the business. They should take expert advice and acquire enough knowledge to cope with business dimensions which are initially unfamiliar. The non-technical manager can learn how design projects are carried out, and become able to relate technical achievement to the commercial objectives of the firm. He can learn at least some of the terminology, to aid communication.

The engineer can quite readily acquire the basics of management, marketing and financial control. It is not necessary to be an expert in these disciplines to make good use of proven techniques. The engineer's inclination to precision is very compatible with the requirements of financial estimation, planning and control. Professor Wayne S. Brown, who has been involved with and responsible for the creation of several successful technology based companies operating in the University of Utah Research Park, is convinced that 'if an engineer is truly interested in learning about accounting, management techniques and business procedures, there is nothing to stop him from being an effective manager' [29].

Product development work

The work involved in product development cannot be divided neatly into the dimensions outlined and illustrated. Practically every task undertaken incorporates aspects of more than one dimension. Recognition of this and attention to the multidimensional aspects will contribute to success. Further discussion of electronic product development work and its management will be conducted with reference to the business dimensions.

References

1 PERRY, T. S.: 'How to make it big: engineers as entrepreneurs', *IEEE Spectrum,* July 1982, p. 55
2 TAYLOR, A. L.: 'Striking it rich', *Time,* Feb. 15 1982, **119,** No. 7, p. 38
3 MAGNET, M.: 'Clive Sinclair's little computer that could', *Fortune,* March 8 1982, **105,** No. 5, p. 78
4 RANDALL, G.: 'Managing new products', Management Survey Report No. 47, British Institute of Management, 1980
5 LANGRISH, J. *et al.*: 'Wealth from knowledge: a study of innovation in industry' (Macmillan, 1972)
6 DRUCKER, P. F.: 'Management' (Pan Books, 1979)
7 GRAY, I.: 'The Engineer in Transition to Management' (IEEE Press, 1979)
8 TWISS, B. C.: 'Managing technological innovation' (Longman, 1980, 2nd edn.)
9 MORRISEY, G.: 'Management by Objectives and Results' (Addison-Wesley, 1970)
10 HUMBLE, J. W.: 'Management by Objectives' (British Institute of Management, 1973)
11 FISHLOCK, D.: 'The Business of Science: Risks and Rewards of Research and Development' (Associated Business Programmes, London, 1975), p. 105
12 SOUDER, W. E.: 'Planning a Career Path from Engineering to Management' (Engineering Management International, 1983)
13 KIDDER, J. T.: 'The Soul of a New Machine' (Little, Brown & Co. Inc., 1981)
14 ROGERS, L. W.: 'The Coming Age of Marketing Maturity' *in* ROGERS, L. W. (ed) 'Marketing Concepts and Strategies in the Next Decade' (Associated Business Programmes, London 1973)
15 WEARDEN, T.: 'Dynamic product strategy', *Electron. & Power,* Nov/Dev 1981, **27,** No. 11, p. 813
16 FRANTZ, G. A., and WIGGENS, R. H.: 'Design case history: Speak and Spell learns to talk', *IEEE Spectrum,* Feb 1982, p. 45

17 BERRIDGE, A. E.: 'Product Innovation and Development' (Business Books, Stockport, 1977)

18 CORFIELD, K. G.: 'Product Design' (NEDO, London, 1979)

19 BYLINSKY, G.: 'Visionary on a Golden Shoestring' *Fortune,* June 1973

20 Science Policy Research Unit, University of Sussex, 'Success and failure in industrial innovation: Report of Project Sappho', Centre for the Study of Industrial Innovation, London, 1972

21 Design Committee of the Institution of Electrical Engineers, submission to the NEDO Study on Product Design, 1978

22 BUGGIE, F. D.: 'Strategies for New Product Development', *Long Range Planning,* 1982, **15,** No. 2, p. 22

23 ROTHWELL, R.: 'From invention to new business via the new venture approach', *Management Decision,* 1975, **13,** No. 1, p. 10

24 JONES, W. S., and PEATTIE, R. C.: 'Micros for Managers' (Peter Peregrinus Ltd., 1981)

25 NORTHCOTT, J., and ROGERS, P.: 'Microelectronics in Industry: What's happening in Britain' (Policy Studies Institute, London, 1982)

26 Computing Services Association 'Applying Microelectronics in Manufacturing Industry — A Guide for Management', Department of Industry MAP Information Centre, 1980.

27 FORESTER, T. (ed): 'The Microelectronics Revolution' (Basil Blackwell, Oxford, 1980)

28 SOUDER, W. E.: 'Effectiveness of Product Development Methods', *Industrial Marketing Management,* 1978, **7,** No. 5

29 BROWN, W. S.: 'University Influence on High Technology Business'. SEFI Conference on The Education of the Engineer for Innovative and Entrepreneurial Activity, Delft University of Technology, June 1982

Chapter 3

Product development process

The electronics industry has a very dynamic image – new products are announced daily, new companies spring up to develop and commercially exploit innovative ideas for products. The business reality behind the image is that electronics companies depend on product development for their creation, growth and survival. New product development is the foundation of the electronics business.

Peter Drucker, the management sage, states that 'the innovative function organises work from where we want to be, back to what we now have to do in order to get there' [1]. In innovative electronics, what has to be done in order to make the business successful is dependent on the product development process. This process is conducted by means of the development project – the set of activities wherein an idea for a new product is transformed by the application of human and material resources into a commercial product suitable for manufacture and sale. Achievement of the objectives of the development project is essential for product success in the marketplace, but it does not guarantee that success. Other factors, such as the effectiveness of production and marketing, financial management and the effect of external economic conditions, contribute to overall business success or failure.

The project can be represented as an activity within the total business (Fig. 3.1). Rather more complicated diagrams, with quantitative indications of expenditure on manpower, materials, services and equipment, could be used to illustrate the relative size or significance of the project, compared to other functions such as production and marketing in a particular company or concern. Just as important are relationships which are time dependent, notably the sequencing of procedures and tasks. In Fig. 3.2 the project is shown as the interface or bridge between product planning and the commercial phase.

Product planning is the process whereby new products are identified and 'adopted' for development. For many companies and concerns, this process forms part of the generation of corporate strategy; for others, the opportunity to develop an attractive new product can effectively dictate corporate planning, particularly where the organisation is small and the degree of innovation is high. The commercial phase encompasses manufacture of the product, its launch on to the market and the development of the sales programme, with associated support activities such as service and maintenance. The commercial phase is undertaken to provide a return on investment and to generate profits for survival, consolidation and further innovation, and to make an adequate contribution to the operating overhead costs of the organisation. To justify continued production and marketing, the new product must pay its way.

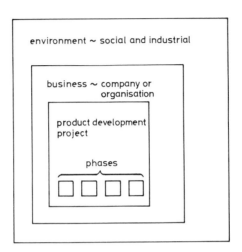

Fig. 3.1.   *Product development within the business*

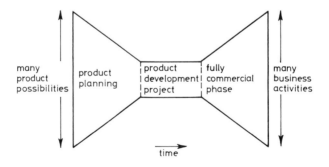

Fig. 3.2.   *Product development in the business sequence*

The axes of Fig. 3.2 are not dimensioned but they serve to indicate that the multiple possibilities which exist during product planning are reduced by elimination and selection of candidates to a product choice for development. On project completion this bridge leads to the many activities required in the commercial phase. The 'butterfly' shape emphasises the central role of product development. Of course most businesses will have more than one butterfly 'in flight', representing products in a range, at different stages of planning, development and exploitation. The butterfly analogy cannot readily be extended into a quantitative model (perhaps fortunately); product planning and commercial exploitation have rather different content and duration, so an attempt at more accurate modelling would produce a somewhat lopsided butterfly, with an inferior flying performance.

Not surprisingly, quantitative representation of the product development project requires the inclusion of financial information. The place of the project in the history of a product, that is relative to the 'product life cycle' depiction so widely used by marketing strategists, can be defined in terms of revenue movement and time.

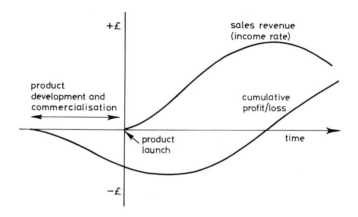

Fig. 3.3. *Product life cycle*

The project and the product life cycle

The product life cycle (Fig. 3.3) is usually represented as starting with the product launch – the introduction to the market place. Sales revenue grows from nothing through an initial acceleration as the market responds, until the rate of sales income decreases and levels off. After this stage, sales revenue in terms of sales income for successive time periods usually falls until the product is withdrawn. In the electronics industry the total time from product market launch to withdrawal can range from several months to several years. The sales revenue pattern can exhibit fluctuations, for example, if production difficulties are encountered, resulting in limited sales and fewer orders taken in a given time period. However, the general shape of the product life cycle graph is as shown in Fig. 3.3 – for successful products. An unsuccessful product may never get into the accelerating sales revenue phase or its sales may collapse later. Even a successful product may be withdrawn while sales are on the plateau or still increasing if commercial considerations dictate that its place should be taken by another. Nevertheless, the product life cycle graph is recognised as a standard representation, particularly by those concerned with new product planning and marketing strategy [2].

To see where the product development project relates to the life cycle, it is necessary to look to the time before product launch. Expenditure on the specific product starts when development work begins, and, for the successful product, continues through the complete life cycle. No return on this investment is possible until after the product has been launched. In addition to design and engineering work, commercialisation must precede launch, as indicated in Fig. 3.3. Commercialisation optimises the product for the market place and puts into place the facilities and procedures for product manufacture, sale and after sales support. The combined demands of development and commercialisation, which can overlap in time, together with inevitable delays in receipts from sales, have the result that the cumulative profit (loss) graph does not reverse its slope until some time after launch. The graph then starts moving towards profitability, with a further delay until adequate sales revenue brings the curve into net profit. At product launch, development and commercialisation costs are usually still being incurred, particularly for the support of early production.

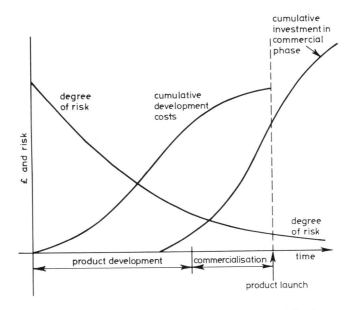

Fig. 3.4.   *Investment and risk for product development and commercialisation*

The pre-launch phase is closely associated with promise; promise that the development project will be successfully completed with the new product ready for manufacture and launch, and promise that this product will be a success in the market place. It is also characterised by revenue movement in one direction — outwards. In compensation, progress on project work should result in a reduction of the associated risk. As the product and the plans for manufacturing and marketing it become better defined, the information on which investment decisions are made is enhanced in accuracy and extent. Better information can and should mean better and less risky decision making.

Fig. 3.4 is a representation of trends in costs and risk before and after product launch. The product development project is bounded by constraints of resources and time. Neither the beginning nor the end of the development work is precisely defined. Product strategy and planning lead into actual development work on the specific product, sometimes with the link of a feasibility study. Commercialisation includes the establishment of suitable manufacturing and marketing facilities and procedures, requiring 'know-how' input from the development phase.

In order to examine the product development process — to be considered in its engineering, management, marketing and money aspects — an understanding of the nature and components of the development project is needed. This must include consideration of the project's interfaces with earlier and later stages of business activity.

The nature of the project

The development project for a new product begins when the product is identified and adopted, and ends when the transition to production is completed. This working definition accommodates 'desk research' at the beginning and 'troubleshooting' at the end. It is not a universally accepted delineation; it can be argued that product development starts when market requirements are formulated, even before the product idea takes shape, and continues through appraisal stages subsequent to launch. Nevertheless, it is logical to concentrate on the period between product definition and product manufacture. This time covers all the activity which is devoted to the individual product, as opposed to product planning concerned with multiple subjects and manufacturing of a range of items. Although progression through the period is accompanied by considerable changes in the nature of the activity and in the staff groupings involved, engineering, marketing, management and money aspects are all significant throughout. The continuity of engineering design work provides a narrative thread.

It is helpful and realistic to consider the product development project as a sequence of stages or phases:

1 Concept and definition.
2 Feasibility study and the business proposal.
3 Design and engineering.
4 Transition to production.

While these concise headings accommodate many activities, the four phases can always be recognised in a development project. At the end of each phase, a decision has to be made — should we go on? Admittedly the room to manoeuvre, in a business sense, is reduced with each phase as commitment is increased. At the same time the risks involved are reduced, as knowledge and experience are enhanced. Not surprisingly, product development project phases are often given other descriptive titles, such as evaluation instead of feasibility study and implementation in place of transfer to production [3, 4]. The process can be further divided, with more than four stages. However, the phases identified here are distinctive and applicable to all electronic product development projects. It is of course the contents of the phases which matter, rather than their names, but an agreed terminology helps to ensure effective communication. This is particularly so within a company or organisation, and agreed 'standardised' definitions and descriptions of stages, activities and functions can avoid misunderstandings and confusions [5]. Of course some product development is pursued, in some companies, with no regard to phase definition. This approach is not recommended.

The sequence of project phases starts with concept and ends with production and these terms refer to very dissimilar activities. As the project progresses, work becomes less 'free style' and more organised; brainstorming sessions can form part of product idea generation, while in contrast the initiation of production requires very precise attention to all the details of manufacturing. Electronic product development can be described as 'task dominant'. As defined by Souder [6], this means that those involved 'have a strong orientation and focus toward the task and the end product; they talk in terms of products rather than functions, objectives rather than processes, task goals rather than functional achievements'. Further task dominant characteristics are that staff are specialists, related to the project 'in terms of their contribution to the team rather than their functions; they have frequent contacts with each other, with crossing and overlapping communication channels'. Readers with experience of product development project work will recognise these features, which relate particularly to the core phase of design and engineering, where authority is determined by expertise as well as position in the managerial hierarchy.

In task dominant project work there is a concentration on the achievement of a series of objectives which become more clearly defined as work proceeds, and the approach adopted to managerial and organisational aspects tends to be adjusted accordingly. In the early stages, flexibility and ease of communication are very important; ideas are welcomed from any source and information is sought in an unrestricted way. This approach has been described as 'organic' [7], and can be implemented by vaguely defined management arrangements. If it is not too contradictory to try to define what is meant by a vaguely defined organisation, the key principle is that each individual professional has a general responsibility to try to achieve the project objectives. 'His authority and tasks are not organisationally defined but are redefined continually through interaction with others.' [8].

Organic systems of management were first identified and examined in depth by Burns and Stalker [7], who studied the organisation of 20 English and Scottish firms, mostly involved in electronics. The work was carried out in the late 1950s and much has changed in electronics since then, but this research has led to a better understanding of factors to be taken into account in the organisation of innovation and development. Other well known work in this area is that of Lawrence and Lorsch in the USA [9].

Burns and Stalker concluded that the organic approach is suitable for innovative work in changing conditions, such as exist in the electronics industry. It is characterised by a 'democratic' atmosphere with great emphasis on professionalism, open communication and flexible task assignment. In contrast, 'mechanistic' management systems are appropriate to stable industrial conditions; there is a clearly defined hierarchic structure of control, authority and communication, with decision making interactions predominantly 'vertical', between superior and subordinate. Of course organic and mechanistic systems are not mutually exclusive; both forms may exist within a company or organisation. They can even be operated by an individual, who may work long and 'unsocial' hours and use equipment and resources 'borrowed' from other departments to achieve design goals in an organic fashion but who will then conform precisely to the mechanistic demands of manufacturing for engineering standards, procedures and documentation related to the product.

The organic approach has been proposed as generally applicable to R & D. However, it is shown by experience that organic management is usually inappropriate to the later stages of product development where well defined structures for production dominate the environment.

A better and more realistic model of the process exhibits a gradual transition from an organic to a well defined mechanistic organisation, as the project progresses from concept to the production stage. Consider the individuals working on a product development. Initially only a few members of staff are involved, preferably representing engineering, marketing, financial and management interests. The prospects for successful innovation are enhanced if no impediments are put in the way of free dialogue and inspiration. Communication is on an informal, 'face-to-face' basis. Of course this freedom is possible only because an adequate infrastructure exists within the company or organisation, coping with more mundane but essential 'day-to-day' activities. The infrastructure is almost certain to be well defined, mechanistic and somewhat rigid. As development proceeds, more individuals become involved and many of them, particularly those concerned with manufacturing and financial control, must have precise information on which to work and make decisions and plans. The project itself adopts more mechanistic features. In addition to changes in organisational methods, leadership characteristics are modified, and the 'bright ideas' individual so important at the product idea stage moves aside for the manager fully conversant with structures and procedures for the commercial phase. In some companies and organisations, the formal mechanistic structures which are key to smooth production, selling and financial control are almost ignored in product development, until work has to start on the transition to production. In others attention must be given to the rigid disciplines of manufacturing, for example, as soon as the product is defined.

A general guideline is that organisational arrangements for product development should allow individuals to contribute their best efforts. This translates into freedom and flexibility to innovate at early stages of the project, and progressively increased attention to precise specification and planning as the product approaches the commercial phase. Details and merits of organic and mechanistic organisational approaches can be debated at length, but product development depends on quality of work as well as organisational efficiency. The organisation of product development has to allow good work to be done at each stage and this implies a changing structure. The natural leader at the concept and definition stage may be a marketing specialist, and may hand over responsibility to an engineer for feasibility work. Continuity can be assured by the first and subsequent leaders providing 'consultancy' input to later stages.

Operating procedures and organisational structures have to be adapted to the different phases of the product development project. They must also be modified with changing conditions in technology and markets. Fortunately, electronic product development usually involves limited numbers of staff on each project and acceptance of change is generally assured, thus facilitating adaptive organisation. The discussion of project phases in the next chapter will indicate organisational features and needs.

Development people

A US company's design team worked 'until 10 o'clock almost every weekday, many Saturdays and some Sundays, for nine frantic months' on a new word processor. Three Northern Ireland university staff each devoted more than 20 hours a week outside normal working hours for over a year to develop a medical data acquisition system. This level of determination and commitment — and more — is not uncommon in the electronic product development business.

In the final analysis the quality of work depends on the quality of the people doing it. Quality includes character, and the work output of highly motivated individuals is many times that of 'time servers'. Management of product development requires the recruitment, stimulation and retention as staff members of qualified and capable professionals who are creative and industrious and prepared to contribute 'above and beyond the call of duty'.

Those involved in electronic project work often represent a number of professional disciplines, with designers, production engineers and marketing specialists most in evidence. A phenomenon of the process is the enthusiasm with which development staff, and design engineers in particular, throw themselves so wholeheartedly into their work. Of course, product design can be very satisfying; after all, engineering has to do with making things.

Successful development depends on human attitudes and teamwork. The zeal and *esprit de corps* displayed in many projects contrast with popular concepts of the sober technologist. Product development can be fun.

The industry and creativity of the rank and file members of the project team must be complemented by effective leadership. Organisation and management of the human resource is the foundation of project planning and control. These topics will receive further examination later.

To complete the general description of the development process, the cast of characters involved will be introduced. In some projects an individual performs more than one role and in others more than one person is required for a single function.

The entrepreneur

This title is conventionally applied to 'an independent, courageous, enthusiastic and tenacious individual who seizes on an idea or invention and who somehow establishes a new enterprise in order to exploit that idea commercially' [10]. The entrepreneurial function is thus often interpreted somewhat narrowly, as confined to the establishment of new small business, and, as such, it has been a subject of much study and optimism for industrial innovation [11]. It is true that setting up a new company is a particularly clear example of entrepreneurial activity, but the entrepreneur exists in many other forms. One such is the 'product champion' within a company or organisation, who promotes a new product. A more general description of the entrepreneur is 'someone who takes the initiative to organise resources in some organisational form for some purpose. The operation has relative autonomy and the person who has thought it up runs it and shares the risk. If the event succeeds, he or she succeeds' [11]. Obviously someone who convinces superiors that a new product should be supported, who argues for resources to be devoted to it, who works enthusiastically on the project in a technical, marketing or management way – in short, who identifies with the product and project – displays most or all of the entrepreneur's characteristics.

The entrepreneurial spirit, whether linked to new business creation or not, is key to success in new product development and often is descriptive of design engineers' and marketeers' attitudes. The entrepreneur may be responsible for one or more professional activities, and a single project may have more than one entre-, preneur committed to it.

Entrepreneurism is a critical requirement for new product success.

Project manager

The first objective of the project manager is the successful completion of the project. Leadership and administration must be provided for the project team. To be effective, the project manager must command the professional respect of subordinates, and an entrepreneurial attitude is highly desirable. The project manager's tasks and responsibilities will be given attention in later chapters.

Design engineer

Engineers are liable to identify the design function as the most important aspect of the project. In the sense that the product actually takes shape in the hands of design staff, this is true, but product design must be directed to satisfying requirements which include market and cost considerations as well as functionality. The design engineer should communicate and be receptive to ideas emanating from other professional disciplines, as well as doing a good technical job.

Production engineer
This is a person to whom the design engineer should pay attention. Production problems can be anticipated and avoided by dialogue on design proposals and often advantage can be taken of new manufacturing methods.

Technological gatekeeper
Referred to extensively in the literature on innovation, this individual acts as an interface to the world of technology outside the organisation. Ideally the gatekeeper attends conferences, reads journals and keeps in touch with professional peers. In this way the project team can be kept up to date with new devices, materials and standards.

Marketeer
This person has been involved in the project since the concept stage. The marketeer keeps the project team 'pointed in the right direction' and can input news of market changes and competitors' efforts.

Controller of resources
Technical and human resources have to be made available to the project in a controlled way. Allocation of resources must be done in the context of overall corporate activity and planning. The controller monitors expenditures and contributes to the business planning function for the new product.

All these characters, who have only been sketched in here, feature in the development process. Chaos would prevail if the differentiated functions listed above were not integrated into a single innovation process. Integration requires communication, within the project team and outside it, particularly to and from the top management of the organisation. The product development process depends on people, their professional skills and their ability to communicate and co-operate.

References

1 DRUCKER, P. F.: 'Management' (Pan Books, 1979)
2 BERRIDGE, A. E.: 'Product Innovation and Development' (Business Books, Stockport 1977)
3 Design Committee of the Institution of Electrical Engineers, submission to the NEDO Study on Product Design, 1978
4 COOPER, R. G.: 'A process model for industrial new product development', *IEEE Trans. Engineering Management*, 1983, **EM–30,** No. 1, p. 2

5 ZOPPOTH, R. C.: 'The use of systems analysis in new product development', *Long Range Planning,* March 1972, p. 23

6 SOUDER, W. E.: 'An exploratory study of the coordinating mechanisms between R & D and marketing as an influence on the innovation process'. Report for the US National Science Foundation, 1977

7 BURNS, T., and STALKER, G. M.: 'The Management of Innovation' (Tavistock, London, 1968)

8 LOVLAND, P.: 'Discussion on principles of organising applied research and development', *Research Policy,* 1974, **2,** Part 4, p. 322

9 LAWRENCE, P. R., and LORSCH, J. W.: 'Organisation and Environment' (Harvard Business Press, 1978)

10 ROTHWELL, R.: 'From invention to new business via the new venture approach', *Management Decision,* 1975, **13,** No. 1, p. 10

11 SWEENEY, G. P.: 'New Entrepreneurship and the Smaller Firm' (Institute for Industrial Research and Standards, Dublin, 1981)

Chapter 4

Project phases

The four constituent phases of product development are:

concept and definition;
feasibility study and the business proposal;
design and engineering;
transition to production.

In Fig. 4.1, the main flow of information through the phases is towards product commercialisation. The subsidiary bidirectional highway accommodates feedback for revision of descriptions, specifications and other product data and for checking back on earlier information and assumptions. The project phases differ in scale and complexity, and, most obviously, in the nature of the work involved. Electronic product development involves many specialised activities unique to the technology together with business methods of wider applicability.

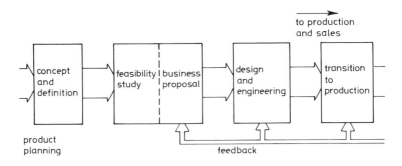

Fig. 4.1. *Project phase sequence*

Concept and definition

In electronics, as in other technologies, product ideas can come from many sources. A company or organisation may have a new-product department, group or officer charged with the task of generating and evaluating product ideas. Engineers working in R & D, salesmen, customers and production staff can contribute to the process [1]. The electronics entrepreneur has a special and important role; thousands of new companies are established, and many of them prosper, because an individual has a novel product concept and the ability and determination to bring it to commercial reality. Many groups and governments have realised that innovation and entrepreneurism are essential for industrial survival and development. Science parks, innovation centres and entrepreneur programmes have been established in many centres in the Western World. Such formalised fostering approaches improve the chances of commercial success for products generated thereby, and there are many indications that such support for innovation and entrepreneurism is well worth while. Institutions of higher education are obviously likely sources of product innovation, since awareness of technological possibilities and comprehensive information resources are assured. Accordingly, universities, polytechnics and colleges are the foci for many initiatives for innovation.

One such American initiative is the Center for Entrepreneurial Development at Carnegie-Mellon University in Pittsburgh [2]. Among the many examples of new businesses established through its work is a taxi company, which uses an innovative electronic meter in its vehicles. The patented meter uses a radio link to a central computer to establish the fare for each journey in advance, to take account of special charges, and to integrate fare revenue into the company's accounts. The equipment was developed by students participating in the Center's graduate programmes.

To give an illustration of such activity in the United Kingdom, the British Technology Group's Academic Enterprise Competition in 1982 had 118 entries from universities and polytechnics. Winners included a group from the University of Southampton which set up a new company manufacturing instrumentation, mainly electronic, for optical fibre communications, and a team from the University of Salford which developed a microprocessor-based system for machinery control.

Consultancy work undertaken by academic staff has often led to the establishment of businesses based on new electronic products or services. Indeed the routes to electronic innovation are legion.

The quest for new product concepts is commonly referred to as product search. Wherever ideas for potential new products come from, searched for or uninvited, it is necessary to subject them to rigorous screening. 'Search' generates product ideas; 'screen' eliminates those which fail to satisfy product criteria, and leads to the selection of candidates for further consideration. The odds against an individual new product idea achieving eventual commercial success are so high that great care must be taken to match commitment to the probability of success. This is not easy. Too stringent and unimaginative an assessment of embryonic new products could eliminate a potential winner; two superficial a screening could lead to substantial expenditure on an eventual failure. From the moment a new product is proposed, effort and money begin to be spent on it. Although the amounts are small at first, they can grow rapidly as development begins and progresses. No organisation can support all the proposals which can be made for expenditure, and new products have great potential for cash consumption. A new product concept may be very promising, getting good assessment from marketing perhaps, but may have to be rejected by a company because its manufacturing facilities are not suitable or are impossible to modify at a cost compatible with commercial prudence. The best that can be done in the evaluation of new product ideas is to try as far as possible to consider all possible and available aspects. A good discipline is to use a set of product criteria (which is kept up to date). The criteria must relate to the company's or organisation's strengths and capabilities. They can then be applied to the new product to determine its suitability for development.

The selection and definition of new products are topics extensively examined by marketing and business planning specialists. Many methodologies and techniques have been proposed and applied. Studies of the processes tend to confirm that effort expended on product selection and definition pays off [3–5]. Although the cost of such work is likely to be just a small proportion of the product's total development cost, the probability of product success can be increased considerably by doing the task really well. Even at this early stage in the product's life, marketing, manufacturing and financial considerations must be addressed. It is unlikely that one individual could adequately select and define a new technological product, although sometimes a winner does emerge that way. Second and third opinions help. The best method is to have a group of specialists, who have detailed knowledge of the strengths and objectives of their company or organisation, the industry and the market, to select, refine and define those product ideas which have the highest potential for success [5]. The new product must satisfy market needs. It must also be compatible with the business environment in which it is to be developed, commercialised, manufactured and sold.

The activities leading to a new product concept and definition can be extensive and time consuming, or concise and rapid. The innovative skills of individuals, the structure and history of the company or concern, the market conditions and the financial climate are all determining factors. Here, interest is confined to electronic products, but such a restriction does not imply a limited range of product cost or complexity, limited markets, or stereotyped organisational structures. The variety of situations ensures a diversity of approach to product planning, leading to the formulation of concepts and definitions. Regardless of the route taken to the point of product definition, a business decision must be made. Does the new product merit further consideration? Obviously, better information will make for better decision making.

Assuming that those concerned are competent, the product definition will have taken into account the known state of the art of the technology, the perceived market need and the conditions existing in the company or organisation. In other words it reflects the knowledge and experience of those responsible. This extends to an appreciation of what is possible; for example, whether custom integrated circuits could be used in the product or automated component insertion equipment could be installed for production. In the electronics environment, the technical possibilities can often be assessed reasonably accurately at the product definition stage, even if quantitative examination is limited to rough sketches and calculations 'on the back of an envelope'.

It is wise to look beyond the limited technical and commercial environment of the company or organisation in question.

What is the competition planning and developing?
What other technical design approaches could be adopted?
Is relevant market research data available from other sources?

The results of a wider look involving informed rough estimates and judgments may lead to modification of the first definition.

To complete the picture, in accordance with emphasis on the business dimensions, management and money aspects have to be considered, at least in outline.

What are the staffing implications?
Are new organisational structures required?
What is the likely cost of technical development?
What would be involved in setting up production, in terms of equipment and expenditure?

The potential benefits should be reviewed.

What are the possible effects on turnover and profitability?
How will the organisation's capabilities and competitiveness be enhanced?
Will the technical development work be eligible for government grant aid?

These questions cannot be answered fully at this stage, but their consideration will provide useful additional information. Experience gained with similar products may make it possible, even at this early stage, to generate outline financial estimates for the project. When such data can be produced economically and quickly, it is clearly sensible to acquire it and use it in the decision-making process.

The output of the concept and definition stage should be a concise and informative document. It should provide sufficient information about the new product to permit engineering, marketing, management and financial interests to appreciate the main features and business implications of the proposed project. It represents a preliminary assessment of the product's commercial and technical viability.

The term 'product definition' as used here should not be confused with the much more detailed product specification required later when design and technical development work commence. At this first concept and definition stage what is needed to justify going further is preliminary data on significant aspects. It is more important that all the business dimensions are considered, using relevant if incomplete information, than that any one aspect, including engineering, is exhaustively researched. Lifetimes of electronic products can be short. The concept and definition work should be carried out as rapidly as possible, consistent with reasonable accuracy. A prolonged market study could end up considering a competitor's product developed in the meantime.

Product concept and definition data may in some instances be enough to justify authorisation of design and engineering work, for example when similar products have been developed successfully already. Even when all indicators are positive, it is more likely that commercial prudence will dictate a further limited commitment of effort, to reduce uncertainty and to verify the product's suitability. This commitment can best take the form of more detailed investigation, including engineering work, particularly for product specification as well as marketing and financial studies. Many product ideas may enter the concept and definition phase, but only relatively few progress to the next stage, feasibility study and the business proposal.

To give an example, a product definition could be in these terms:

General description: An electronic lock for suitcases and similar items, to replace mechanical combination locks.
Features: Keyboard control, battery powered, size less than limits (specified). Failsafe. Suitable for automated manufacture.
Manufacturing cost: Less than £X per unit for production rate (specified).
Market: Manufacturers of quality suitcases, briefcases etc. Quantity sales, on predetermined call off basis.
Development programme: To be completed by (date) to accommodate product announcement and sales programme.
Organisation for manufacture and sale: The proposed structures for management and workforce, or guidelines to be followed, would be noted here.
Finance: Indications of required relationships between manufacturing cost, selling price and product volumes would be noted here.

Supporting material could include preliminary market data, artist's renderings and notes on technology.

It is unlikely that a definition of this kind would be adequate to justify the initiation of a full technical and marketing development programme. But supported by information on other factors relevant to this product and to the company which is considering its adoption, such as the available experience of related products, technology and markets and preliminary financial data, the definition, if sufficiently attractive, allows a further step to be taken towards product realisation. In effect, what has been achieved so far is an acceptance in principle that the product satisfies, in general terms, the required criteria, and that further limited expenditure on it is merited. The product's feasibility must now be assessed, and, depending on the result, consideration given to preparation of a business proposal.

Feasibility study and the business proposal

The time for rough sketches, jottings, rapid 'back of an envelope' calculations and 'ball park' estimates is not yet over, but those engaged in feasibility study have more to 'get their teeth into' than the product planners. A specific new product has been defined and now has to be assessed technically and commercially. The assessment and the technical and business development which may follow could of course cause the definition to be altered. Indeed, during the transition from definition to technical specification, changes are practically inevitable, for example in product features and operation. The product definition as discussed is nonetheless the starting point for really quantitative work. This work must provide answers to four questions:

1 Can the product be sold?
2 Can the product be designed and engineered?
3 Can the product be manufactured?
4 Does the business proposal make commerical sense?

These questions are interlinked. For example, the sales prospects in volume terms may be very impressive, but high development costs coupled with limited profit margins could render a business proposal based on the new product unacceptable. Furthermore, the only practical way to answer the questions requires examination of what is to be done, why, when and how. There will usually be scope for flexibility in arrangements for development, manufacture and sale, including the use of in-house and subcontracted resources. These options have to be considered. Obviously, the better the concept and definition phase has been carried out, the more specific can be the work on feasibility. This work has to be quantitative and provides a product specification, detailed estimates of human, physical and financial resources needed and proposals for how they can best be used. Sometimes feasibility studies are undertaken with limited objectives; for example, to establish if microprocessor technology should be adopted and to specify the most suitable devices. It is usually more meaningful to have the feasibility study work addressed to all aspects of the product. The resulting information, if sufficiently encouraging, makes possible the formulation of a business proposal. On the basis of this proposal, a decision can be made on whether or not to initiate design and engineering work and to plan for production and sales.

Feasibility assessment and the associated business proposal (for a promising product) are key aspects of product development. If the decision is subsequently taken to proceed with product design and commercial development, reports produced at this stage form the technical and business foundations for the ongoing work. The next two chapters are devoted to feasibility study and the business proposal. Before going on here to outline the next stage of development, which is design and engineering, some general points can be made about essential aspects of feasibility study.

The feasibility study has to be carried out efficiently and competently. This means that experienced professional staff should be responsible. If specialist expertise, for example in such topics as microprocessor firmware or leadless components, is needed and is not available in-house, suitable outside consultancy services should be utilised. There are many topics to be examined and check lists are helpful. Good documentation is absolutely essential, with unambiguous conclusions and well defined plans. It is often useful to carry out limited practical work or computer simulation to investigate specific options. Preferably, one staff member should have overall responsibility for this feasibility stage, liaising with specialists in engineering, marketing and finance. Feasibility means fine detail.

In many ways the feasibility study is a project in itself. Its output is a set of reports and proposals, which is of less interest to engineers and marketing staff than a real new product in prototype or preproduction form would be. For this reason, and because the work is so important, top management must pay special attention to ensuring that the work is given appropriate priority and status. Feasibility study must not be a background activity; the new product's momentum must be maintained.

At this stage of the product's progress, rigid organisational structures are inappropriate. Departmental divisions must not inhibit free ranging dialogue between engineering, marketing and finance. Leadership can be provided successfully by either engineering or marketing, since both disciplines focus on the product. It is helpful for engineering to be reminded of the market requirements for the product, while marketing people should appreciate technical possibilities and limitations. Both groups need to be cost conscious.

While a model approach to feasibility assessment can be helpful, it is clearly restrictive to field the same team and the same technique for every new product. Engineers can transfer from design and technical development duties to feasibility work to ensure that up to date technical information is being used. Marketing specialists who stay involved with the market place can contribute most effectively to new product assessment.

An effective approach is to form a temporary product group, tasked with carrying out the feasibility study and generating the resulting business proposal if this is justified. One or two individuals can be assigned full time, perhaps for a month, while others can contribute on a part time basis. Such a temporary grouping policy enhances flexibility and establishes an informal and effective communication network within the organisation. This network can be very useful during the next stage, which is product design and engineering.

The feasibility stage can be truly innovative, and it is best if originality can be encouraged, so that the proposals for further work are comprehensive. In other words, the assessment of the new product should be carried out with a positive attitude, while keeping eyes open for possible obstacles to technical development, manufacturing and selling. The resulting business proposal can thus provide a reliable evaluation of the project, including the first realistic financial analysis.

The feasibility study is a means of estimating the product's chances of technical and commercial success — the business proposal is a mechanism to justify and obtain financial support for the product's further development. The project phase which accommodates them requires effort on all the development dimensions.

Design and engineering

This is the most visible phase. It begins with information; analyses, reports, specifications and plans — and should end with the product as a physical entity, ready for production. Sometimes engineers and others regard this phase as the totality of product development, but such an attitude is dangerous. Success is more likely if the preparatory work preceding design and the early stages of production following it are given the same painstaking attention that is needed for technical development.

Depending on the product, 2, 20 or even more individuals contribute to this phase, for 2, 20 or more months. The cost of manpower, materials, services and resource utilisation is never trivial, while the cost consequences of design and engineering decisions for later commercial stages are always critical. In many ways this phase generates the most stringent demands for planning and control functions. These topics are addressed in Chapter 7. Here, discussion will be concentrated on what is involved in product design and engineering, without going into technical detail. For many engineers, it is the technical detail that counts, but that is the province of other books.

Technological environment

Electronic product design and engineering techniques are constantly changing as new component technologies and new equipment become available. A principal engine of change has been the integrated circuit industry, with the rising level of integration in solid-state components, bringing with it a corresponding increase in product and system complexity. In the 1960s, products based on discrete components and resistor-transistor logic gates each commonly utilised 1 to 10 000 active devices, or transistor equivalents. In the 1980s, the figure is in the region of 10 to 100 million transistors per system, as in, for example, a personal computer. The increase in devices accommodated per chip is exemplified by progress in random access memories, with only 6 years spanning the launches of 16 K, 64 K and 256 K parts.

During the two decades from 1960 the complexity of integrated circuits available commercially maintained a growth rate of about 100% per year. This doubling of complexity each year is often illustrated by a plot of the logarithm of number of devices per chip against time of availability (Fig. 4.2). During the same period, chip manufacturing improvements gave dramatic cost benefits, so that the real price of a logic function was roughly halved each year. These developments alone would have ensured that 'technology push' generated exciting new electronic products, and of course it did, with pocket calculators, digital watches and many others, but there was more to come; the stored program concept confined heretofore to mainframe and minicomputers became available for electronic products and systems in general with the mid-1970s advent of the microprocessor, memory and support chip families. Electronics engineers added software, chip embedded firmware and computer architecture development to their set of skills. Earlier reliance on logic and circuit design was replaced by a much broader approach, extending from knowledge of microelectronics fabrication possibilities and limitations to familiarity with complete computer system design.

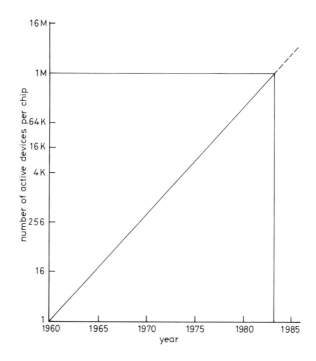

Fig. 4.2.   *Chip complexity history*

Just a few decades ago it was unimaginable that products of the complexity and power common today could be made available to customers at present prices. The modern electronic product is typically characterised by sophisticated design, complex devices and automated manufacture. Assembly, which only a few years earlier represented up to 50% of the cost of manufacturing, was reduced to 20% or less by 1980 for products built in substantial volume. The modern electronic factory uses program controlled component preparation, automatic component insertion, computer controlled soldering, testing and fault finding. Industry standards, notably for printed circuit board assemblies, are in many aspects applicable not only to large scale and highly automated manufacture but also to small volume, manual production. The way in which a product is to be manufactured has to be carefully considered at the design stage, for example, to ensure that automatic component insertion equipment can be used.

All these factors – large scale integration, software and firmware, the potential or need for automated production – make the design process, as well as the product itself, complex. Design costs increase with complexity, but fortunately not at the same rate. Design aids can use the same new technology that provides for the new products. Fig. 4.3 indicates progress in design equipment with time and integrated circuit complexity. The availability of computer-aided development systems has helped to contain design cost escalation, but availability does not ensure application, and much 'manual' design continues. The general picture (Fig. 4.4) is one of lengthening product development time, while product lifetimes decrease, particularly for consumer electronics. It is commercially dangerous to have to embark upon and make substantial investment in the technical development of second-generation products before the first generation is even starting to pay its way in the market-place. The conflict between product development time and product lifetime, which has been referred to as a 'crisis in design productivity' [6] can be resolved by utilising modern computer-aided techniques to speed and enhance design.

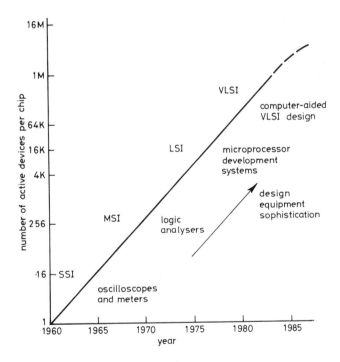

Fig. 4.3. *Design equipment and chip complexity*

At the present time (1984), electronic products are predominantly printed-circuit-board (PCB) based with either one or two PCBs mounted in a specially designed mechanical enclosure, or a set of PCBs in a rack or frame housed in an instrumentation case. Multilayer and flexible PCBs may be used for special applications, but the great majority of products use double-sided boards, with plated through holes to provide connections between the sides. Integrated circuit packages 'stuffed' onto the PCBs are defined as incorporating small (SSI), medium (MSI) or large scale (LSI) integration. SSI means 10 to 100 devices per chip, MSI means 100 to 1000 and LSI means 1000 to over 100 000. Most PCBs carry at least a few discrete components such as resistors, capacitors or diodes.

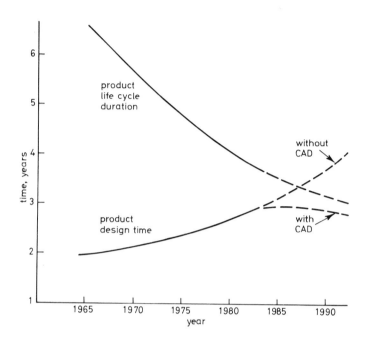

Fig. 4.4.   *Product lifetime and product design time: the effect of computer-aided design (CAD)*

It has been estimated that about 95% of the logic functions on a typical digital product's PCB are provided by a few LSI chips (microprocessor, memory and input/output). The remaining 5% is performed by supporter 'glue' chips which occupy about 65% of the PCB space and dissipate a similar proportion of the electrical power used [7]. The cost effectiveness of the complex chips is reflected in the fact that the sale of support components is worth, in money terms, three to four times the value of the LSI devices which provide the lion's share of processing power. Analogue circuits have not lent themselves to integration as intense as digital logic. PCBs in analogue products are heavily populated with the ubiquitous integrated operational amplifier and its relatives, together with several discrete components for each chip.

Digital products and systems are heavily dependent on firmware, imprinted on read-only memory. More than 80% of the added value (sale price less the cost of brought in materials and components) can be provided by the firmware.

The electronics designer is not limited to the use of standard 'off the shelf' items. Integrated circuits to a unique functional and design specification can be 'customised'. Such custom chips can be cost effective where no suitable standard components exist or where standard microprocessor/memory configurations mean uneconomic design 'overkills'. Depending on the technique used, a customised integrated circuit can have a few hundred to several thousand active devices on board. Many microelectronic manufacturers offer such circuits and in some cases the customer − the design engineer − can undertake a considerable part of the chip design work personally. Many new electronic products contain at least one custom integrated circuit. For example, an LSI custom chip is the real heart of a teddy bear with an audible heart beat, designed as a consumer product to comfort babies away from their mothers [8]. Its Florida originators, Rock-A-Bye Baby Inc., adopted a custom chip to save space − without it 'the product would have had to be a small gorilla rather than a teddy bear', in the words of consulting engineer Barry Greenberg.

Fig. 4.5. *Manual production of PCB artwork*

Beyond LSI is, naturally enough, VLSI, with hundreds of thousands of devices on a chip; for example, a Hewlett-Packard 32 bit microprocessor contains over 450 000 transistor equivalents. The number of possible useful standard VLSI parts is limited; as building blocks they can be functionally too large, but a VLSI chip can have capacity for all the functions needed for a specific application, and product design then becomes mostly custom VLSI design. Such advances are already altering many aspects of electronic product development, but design is still (in 1984) predominantly at the system level, using standard parts and some custom types mounted on PCBs. Figs. 4.5 to 4.7 illustrate electronic design activity.

Computer-aided procedures are important and often essential for design and maufacture. Circuits can be designed and verified by simulation, thus reducing the need for 'breadboard' prototype versions. PCB artwork can be prepared accurately and rapidly. Manufacturing data can be generated for PCB fabrication

Fig. 4.6.   *Computer-aided electronic design*

Fig. 4.7.   *Design of ULA custom integrated circuits*

and drilling, automatic component insertion, product assembly and automated testing, and very significantly, project planning, control and documentation can be integrated with the design process.

As always, electronic product design is in transition, but independent of whether manual or computer aided procedures are used, design objectives of functionality, performance, cost effectiveness, ease of manufacture, reliability, maintainability and customer appeal must be pursued.

Design factors

In electronics, design means detail. There are also many important general considerations. The following is a list of product factors to be considered:

Functional specification
Performance margins
Ergonomics
Aesthetics
Customer appeal
Ease and economy of manufacture and test
Packaging
Safety
Reliability
Serviceability and spares
Compatibility (with other products and systems)
Standards and certification
Product lifetime
Sales support
Product derivatives

Each of these aspects requires an amount of work and some products involve additional factors.

Design has been defined as 'the co-ordinated activities of a number of specialists to achieve an acceptable solution to a requirement specification' [9]. Co-ordination means management of staff and resources, requirement implies market need and a specification is an essential prerequisite.

Product specification

The starting point for design work is the product specification. This must be comprehensive and detailed; it is always tempting to leave a gap in the specification 'because this is a development project after all', but such an approach can bring eventual disaster. Thorough feasibility study should have removed uncertainties in the proposed product characteristics and in the design methodology to be adopted. Lack of precision in the specification in many cases means lack of knowledge. Any 'open' features which have to be left in the specification have implications which should be considered and corresponding options should be defined. The resources of time and money needed to implement alternatives are unlikely to be identical, thus complicating project planning. While every product's specification is important for design to be undertaken, those products which interface with other systems, or which must conform to particular standards because of their application, need particularly precise definition.

Important aspects of the specification are:

Overall product objectives: this document defines the product's function and proposed application and specifies the performance characteristics to be achieved. Such parameters as data handling capability, dynamic range, power consumption and interfacing standards are quantified (depending on their relevance to the product). Reliability, environmental constraints and manufacturing techniques to be adopted are included.

Design approach: how is the product functionality to be divided between electronic hardware, firmware and software? Having broken down the product's functions as precisely as possible, an approach to design must be determined for each identified module or subsystem. Choices of integrated circuit technology, microprocessor type, approach to firmware development, physical implementation of circuitry and other technical aspects have to be decided. Design standards to be adopted, assembly procedures and product testing requirements are listed.

Documentation: the written specification sets the scene for project documentation; if a good start is made with attention paid to precision and clarity it is more likely that subsequently produced reports, assembly and test procedures and operating and service manuals will be to professional standards.

Team and tasks

The skills and technical resources for the design and engineering phase are now clear. In addition, a plan is needed, but first the project team must be formed. The team leader or project manager is appointed, and preferably makes the major contribution to staff selection. Individuals may be assigned to the project for its duration or may have part of their time allocated to it. As to team size, this depends on the project. Many microprocessor-based products have been successfully designed and engineered by two or three professionals, with hardware and software development tasks shared between them, using a minimum of formal organisation. More manpower does not guarantee better development achievement but some projects have so many design and engineering aspects that a formalised team structure is essential. Key team disciplines are:

Electronic hardware
Software and firmware
Production
Human factors/ergonomics/mechanical
Service
Administration/documentation

The project manager presides over all. Most team structures have one or more of the disciplinary responsibilities above assigned to individuals.

The project manager may have been closely involved in the preparation of the specification or may have to become familiar with it; in any case he or she must plan to match resources to need to achieve the specified objectives. It may be necessary to arrange access to expertise outside the company or organisation and the monitoring and control of such services need special attention. Project planning is a substantial topic in itself, but a concise check list can indicate key features. The plan should contain:

Project targets and milestones (specific technical achievements and dates)
Allocation of manpower and technical resources
Cost analyses and budget
Project staff structure
Monitoring and reporting procedures
Documentation schedule
Training arrangements (where necessary)

A review procedure involving top management is essential. Development work can be very expensive and a return on the investment is at best some way ahead. Commercial decisions should be made at predetermined milestones of achievement or of elapsed time and expenditure to continue or to stop the project. It is quite likely that design work or market intelligence gathering will cause changes in the product specification and the project plan. These can be incorporated as updates. Rational use of new information in a methodical way is preferable to insular adherence to every detail of a plan which is no longer totally relevant. This does not devalue planning; it is much easier to incorporate defined changes in a good plan than it is to make an inadequate plan workable by drastic and belated modification.

Practical procedures
At the core of the project is the physical realisation of the product, first perhaps as an experimental model, then as a prototype, then in a form for transfer to full scale production and finally as a production item incorporating all the lessons learned in the process. The use of computer-aided engineering can expedite dramatically the practical realisation of the final product.

The practical steps involved are not rigidly defined. Circuits can be designed on a scrap of paper or using the visual display of a computer-based engineering work station. 'Breadboard' circuits can be 'lashed up' or logic designs can be verified by computer simulation. Connections can be soldered or wirewrapped, prototype PCBs can use wire links between sides or through hole plating.

'Traditionally', design work outside the larger companies is manual. Circuits sketched roughly are verified by building and 'debugging' breadboard hardware. Circuit schematics to documentation standards are then prepared by the engineer himself or by a draughtsman. PCB artwork is produced manually, prototype PCBs are commissioned, and final circuits are verified and tested. This approach is widely used by individuals and small development groups. It can be cost effective, particularly if familiar circuit techniques are used. Many PCB companies provide fast prototype board turnaround and a complete product can be assembled in a standard rack or enclosure. It is not uncommon to have quite respectable looking prototype units of simple products within a month of design work commencing, but how many corners have been cut to get there? Marginal design may not show up until production and sales are well underway. A first prototype which works well is very encouraging but the product may not survive the transition to production. Attention must be given to reliability in anticipated environmental conditions, cost effectiveness of manufacture and the other design factors.

Selection of components and accessories is critical. Second sources should be available for all standard items, and steps must be taken to assure as far as possible supply of special components used, such as custom chips. Lead times for component availability can jeopardise progress to volume production.

In the many cases where the product is microprocessor based, firmware is often more demanding than hardware. This is the case where a standard or near standard microcomputer structure is adopted. Sophisticated (and expensive) microprocessor development systems can be used but a great deal of work can also be done using assembler packages on personal computers. Again, even a small team can produce results very rapidly. The concurrent risks of inadequate or non-existent documentation are considerable. Nevertheless, the small entrepreneurial group can be highly productive and its product can complete the course successfully if proper care and attention are given to all the details, mundane and otherwise.

Every development team, large or small, well equipped or not, is vulnerable to the misconception that a first working prototype represents complete success. The milestone is significant, but a prototype is not a production model. Typically, a prototype is available at a point when expenditure of money and time has reached 30% to 40% of the final design figures. Prototype evaluation and incorporation of essential changes into the design for full production has to be painstaking. It can be tedious, time consuming and expensive, but it is much preferable to production started with an inadequate design and engineering specification. Errors detected at the production stage can generate costs dwarfing the expenditure which would have put things right in the design phase.

Computer-aided design, manufacture and test (CADMAT) changes dramatically the nature of practical design and product engineering. It can be applied to all aspects of electronic product design; chips, circuits, layout, interconnections, assemblies and enclosures. Minicomputer-based systems were available from the early 1970s, initially applied to draughting and documentation, with later extension to PCB and LSI chip layout functions [10]. The microprocessor-based systems, first appearing in 1980, bring the CAD of CADMAT within the reach of the small company. Key functions are:

Data retrieval
Schematic circuit design data capture
Logic simulation
PCB design data capture
Data generation for manufacture (PCB drilling schedules, control data for automatic component insertion and product testing, for example)

Firmware design
Report and documentation generation

A notable advantage of CAD implemented via an engineering workstation is that routine and essential tasks unpopular with engineers are readily and relatively painlessly undertaken. These tasks include preparation of parts lists, component source data, engineering change records and general reports and documentation. The CAD system's database helps to ensure accuracy of information used in design. For PCB layout, reduction in design time by a factor of three to four over manual methods is readily achieved. The benefits of uniformity provided are valuable in manufacturing. The hand wired breadboard can be dispensed with, since the circuit design is simulated and verified. CAD systems can provide help in project control by ensuring compatibility between designs being generated by different engineers.

The productivity of the individual design engineer is enhanced by the use of CAD. The project manager can integrate project record keeping and documentation with the design process.

Engineering aspects
Design tasks dominate the early stages of technical development work and (usually) attention is concentrated on implementing a working prototype. From then on, engineering aspects are in the ascendant. Innovatory design may have made the product possible; attention to engineering can make it a commercial success. Some practical notes follow:

Standards: make sure the product conforms to legally enforceable requirements and consider professional standards which are relevant. Adherence to published criteria can expedite testing and customer acceptance. For example, if PCBs can conform to a widely accepted format such as 'Eurocard', many compatible accessories and enclosures can be utilised.
Power supplies: often given inadequate attention. Use well tried and tested circuits if possible, with conventional voltages. If the system is mains driven, cooling arrangements are important, possibly using a fan.
Enclosures: this is what the customer sees. Circuit design engineers are not usually at their best in choosing the product presentation. Rack-based systems can use one of the many attractive instrumentation cases available from vendors; custom designed cases are likely to be expensive and difficult to get right the first time around.

Electronic products, particularly those for consumer markets, often require specialised housings. Shape, size and material are just some of the factors to be considered. The interface between the product and its human user is critical and is best taken care of by qualified and experienced 'human factors' specialists. Ergonomic design should proceed in synchronism with electronic development. Constraints on physical aspects of the product enclosure often affect circuitry, including component choice and layout.

A cost effective approach to product presentation can be to engage a professional, specialist organisation. Its industrial design work involves procedures which are outside the professional expertise and experience of the electronic engineer. Consideration of human factors, safety, convenience in use, visual and expressive values should be undertaken early in the design and engineering phase.

Decision time

The design and engineering phase ends with a review, preferably involving the contributors to this phase and top management. Prototypes or preproduction models (similar to prototypes in specification but built by the manufacturing department), documentary evidence of the product's status and capability and comprehensive reports on manufacturing, marketing and financial aspects are presented and examined. If it becomes clear that further questions must be answered before a decision on production launch is possible, steps are taken to acquire the needed information.

At this point, a substantial investment has been made in the project. This alone does not justify going on into production and market launch. What has been spent is effectively gone for ever; the decision to proceed or not has to be made on the basis of the investment now needed, the probable return on that investment and the ability to afford the risk of the worst case loss. Commitment to the product must be tempered by commercial prudence – the company itself may be at stake.

Transition to production

In the well run project, formal and informal communications between the production department and the technical development team are established at the feasibility phase and maintained throughout design and engineering. Accurate analyses of manufacturing aspects form part of the product database. In the wake of the decision to proceed into production, yet more detailed planning is required. Production people move to centre stage but the technical development staff still have major roles to play.

A change in component value, movement of a PCB track or reorganisation of interboard wiring to improve the product can be trivial to the design engineer. Such modifications can stop a production line. Uncertainty has no place in manufacturing. The production manager is concerned with component acceptance rates, assembly throughput and product quality. Procedures have to be mechanistic and well documented.

The transition to production is certain to encounter difficulties but the negative consequences can be minimised. First, much useful information can be obtained by having several new product units built and tested by production staff. To do this, such items as parts lists, assembly schedules, drawings and test procedures have to be available. In the light of feedback from this experience, formalised production documentation can be finalised. Production jigs, fixtures and tooling are prepared. Purchasing arrangements are made. The details of what is needed are a function of the organisation and of its methods but again the test for acceptability is effective utilisation of the material without intervention by technical development staff who know 'what not to do'. Sometimes subcontracted manufacture is used for part or all of the product. This requires special care since geographic separation can exacerbate communication difficulties and it always takes longer and costs more to correct errors when they occur further away. Commitment to remote manufacture without a trial production run in house is really living dangerously. Indeed, there is great benefit from carrying out a pilot production exercise, perhaps to build 50 or 100 units for promotion and test marketing, even if full scale production is to take place at the same location. Experience of procedures, analyses of time taken for component preparation, assembly, test and fault finding with even a relatively small production sample can bring into the open previously unseen hazards.

The phase of transition into manufacture can also be termed the qualification and pilot production stage. Depending on the company and its market, pilot quantities can be 100, 1000 or even more. What is important is that this last opportunity is taken to get product design and manufacturing as correct and optimum as possible. A pause to consider all the facts before full production is started is helpful.

At this time, product warranty and service procedures are defined and assessed. Any necessary revisions of the overall business plan for the new product are made.

In an ideal world, the design engineer could now turn his or her attention to other projects and leave the new product in the competent care of production, sales and servicing colleagues. In the real world, this may happen, but not for long. Queries from production on assembly and test procedures, requests for component changes from purchasing and visits to customers for sales staff combine to generate a considerable post-development work load. Management should recognise that these occurrences are inevitable and allocate engineering staff time to cope. Again, formalised and documented procedures are valuable, so that lessons learned are retained and staff involvement is defined and controlled. The 'engineering change order' or 'change note' is an effective mechanism to ensure precise definition of a technical modification and provide proper approval by responsible staff.

Even large and well run organisations tend to ignore the post-development engineering work load. Sometimes the only way a development project team leader or key member can be rescued from an indefinite sentence of problem solving for past products is by a move to another post! In the small company, the apparently never-ending responsibility for many aspects of a growing number of products can be demoralising. This problem can be overcome by training, staff deployment and good documentation. A formalised 'handover to production' procedure is essential.

Comment on project phases

This has been a rapid transit of the phases in a product development project. Just as the project is the core of business development, so those activities involving design and engineering, with related marketing, management and financial functions, form the heart of the project. Feasibility study and the business proposal, and the design and engineering phase represent a focusing of talent and resources on the new product. In the following chapters we will look more closely at these central phases.

References

1 BERRIDGE, A. E.: 'Product Innovation and Development' (Business Books, Stockport, 1977)
2 WOLKOMIR, R.: 'He tells students to go and make a business happen', *Smithsonian,* January 1983
3 MATHOT, G. B. M.: 'How to get new products to market quicker', *Long Range Planning,* 1982, **15,** No. 6, p. 20
4 CANNON, T.: 'New product development', *Euro. J. Marketing,* **12,** No. 3, p. 215
5 BUGGIE, F. D.: 'Strategies for new product development', *Long Range Planning,* 1982, **15,** No. 2, p. 22
6 BERESFORD, R.: 'Engineering work stations complete the network of design-automation tools', *Electronics,* 1982, **55,** No. 23, p. 123
7 MEAD, C. A., and LEWICKI, G.: 'Silicon compilers and foundries will usher in user-designed VLSI', *Electronics,* 1982, **55,** No. 16
8 MATTERA, P.: 'The boom in tailor-made chips', *Fortune,* 1981, **103,** No. 5, p. 122
9 Design Committee of the Institution of Electrical Engineers, submission to the NEDO Study on Product Design, 1978
10 HILLIER, W. E.: 'The computer-aided engineering workstation', *Electron. Power,* 1983, **29,** No. 1, p. 69

Chapter 5

Feasibility study

Can the product be sold?
Can the product be designed and engineered?
Can the product be manufactured?

If the answer to each of these questions is yes, the product has a real chance of success. To reach the yes/no decision points requires the acquisition of realistic and reasonably accurate data. The feasibility study and the business proposal which is subsequently prepared for the product with promise provide a bridging structure between the predominantly qualitative activities of the product concept and definition phase and precisely quantitative design and engineering activity.

Verification that the product can generate a sufficient level of sales, that it can be designed, engineered and manufactured, and that all these processes can be carried out at acceptable cost, represent a positive conclusion to the feasibility study. This being so, an appropriate business plan and business proposal can then be formulated. The decision to proceed with substantial expenditure on design and engineering can only be justified on commercial grounds. Concurrent with the assessment of viability, the definitive product specification is generated.

Feasibility

The concept and definition phase has provided an initial description of the product and indications of who it will be sold to and how it will relate to market segments and competitors' wares. It is now necessary to establish if the product can be designed and manufactured to this description and requirement, or to modified but acceptable versions, and if it has enough market potential to justify investment in design, engineering and commercialisation.

A technically excellent product with no prospect of sales is commercially useless. Alternatively, there may be a well defined market requirement for a product which cannot be developed or manufactured by a particular company — and it is fruitless for the company to invest time and money in that unattainable product. A new product must be feasible in both marketing and technical terms. Accordingly, the investigation of feasibility must proceed on the two equally important fronts. There is scope for debate as to which activity should be initiated first, but most professional observers agree that marketing and technical assessments should be carried out concurrently. This allows for rapid response to changing commercial needs, limitations and opportunities, and facilitates the establishment of good communications between practitioners of marketing, engineering and management skills.

The electronics industry does not, in general, suffer too badly from conflicts between engineering and marketing 'factions'. Perhaps this is because both sides use a common vocabulary; many electronic consumer products are sold 'technically' and products for industry have to be promoted for their specialist markets in engineering terms. Whatever the reason — and the relative youth of the industry is surely a factor — the twin tasks of marketing and technical feasibility assessments are usually undertaken with enthusiasm and perception. They may not always be completed with the same quality of attitude. Engineering staff are often tempted to 'lash up a prototype' to see if it can be made to work, and enthused by the challenges of design and implementation, may ignore detailed market requirements. Marketing staff may become frustrated at the engineers' apparent retreat into technical obscurities. Engineers may underestimate the importance of the marketing dimension and denigrate market research and study. Marketing specialists may become disappointed at engineers' lack of commercial acuity. It is a management responsibility to overcome such failures of communication and appreciation and to keep feasibility work pointed in the right direction and properly balanced.

The overall objective of a feasibility study is to acquire sufficient information, of sufficient quality and reliability, on which to base investment decisions. The key word is sufficient — if enough has been done to demonstrate clearly that the new product can be implemented and that it has a high probability of commercial success, effort should be directed immediately to the preparation of the business proposal. On the other hand, if it becomes clear that the commercial risks are unacceptably high, the product should be terminated and attention given to other contenders. Judgment of the new product's chance of success and risk of failure is in the final analysis a function of the company's trading and resource situation. Various methods to assist the judgment process have been used and studied [1]. They include check lists of evaluation criteria, listing many factors [2] under such section headings as product features, marketing, technical development, production, financial and corporate policy. An extension of the check list is the product profile chart [3, 4] where a score is given for each factor and the results are presented graphically (Fig. 5.1). These formalised procedures help to ensure that essential criteria are not overlooked but their inflexible application can obscure important detail. No formula-based analysis can take the place of first-hand knowledge of technology and market-place. The individual who is responsible for managing the feasibility study must apply acquired information and experience as effectively as possible to form a judgment. It is clearly sensible to use any check lists, *aides-mémoire* and profile charts which are found to be helpful, but precise data, awareness and experience are the essentials for accurate product assessment.

The information base to be established by the feasibility study can be defined in terms of criteria for evaluation. The following list of factors includes most of

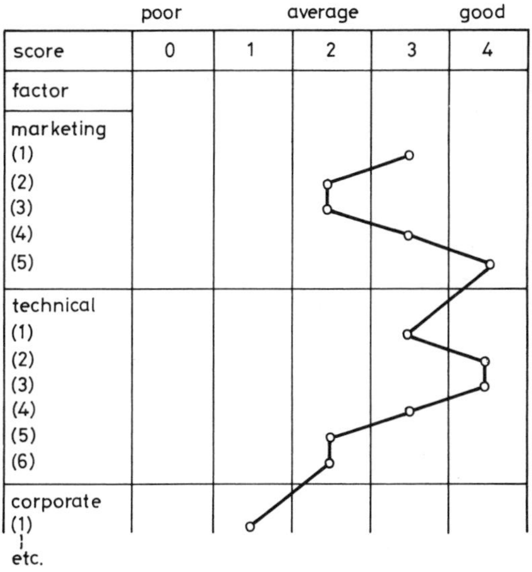

Fig. 5.1. *Product profile*

those relevant in practice. Not all of them are applicable to every new electronic product and they are not of equal importance. In fact the relative importance of the criteria is a function of the particular commercial organisation's interests and strengths.

A Product
  1 Functional and performance specifications
  2 Electronic hardware, software and/or firmware content
  3 Packaging and presentation
  4 Warranty and maintenance requirements
  5 Design protection (patenting or design registration)
  6 Derivatives

B Marketing
  1 Verification of market need
  2 Estimated market size and geographic distribution
  3 Estimated market share
  4 Estimated product lifetime

    5 Probable sales volume and distribution with time
    6 Estimated selling price
    7 Present and potential competition
    8 Additional market research requirements
    9 Test market proposals
   10 Market launch costs
   11 Distribution and selling channels

C Technical development (design and engineering)
    1 Verification that technical development is achievable
    2 Existing resources and know-how
    3 Design approach
    4 Additional skills and resources needed
    5 Anticipated design and engineering difficulties
    6 Estimated structure of technical development programme
    7 Management of technical development programme
    8 Budget for technical development
    9 Potential for further development
   10 Relationship with other projects
   11 Documentation requirements
   12 Applicable standards

D Production
    1 Proposed production methods
    2 Existing manufacturing resources
    3 Additional skills and resources needed
    4 Staff training needs
    5 Pilot production plans
    6 Material and component sourcing
    7 Assembly, testing and fault-finding procedures
    8 Production management
    9 Sub-contracting aspects
   10 Full scale production arrangements

E Financial
    1 Cost of technical development programme (capital and revenue)
    2 Investment in market research and marketing
    3 Investment in manufacturing
    4 Manufacturing cost analysis per unit, i.e. materials and components, direct labour, added value, indirect costs, 'factory door' price, 'break-even' analysis
    5 Cash flow predictions
    6 Probable profitability
    7 Availability of finance
    8 Eligibility for grant aid

These financial factors provide the basis for the business proposal which follows the feasibility study of a promising product. As feasibility criteria they contribute to the go/no go decision making process; for example, inadequate profitability or excessive investment requirement would indicate rejection of the product at this stage.

F Corporate policy
1 Does the project 'fit into' the company's manufacturing and marketing strategies?
2 Do the required investment and estimated profitability comply with the company's policy on risk taking?
3 Are the proposed organisational changes and resource acquisitions consistent with the company's development/expansion plans?
4 Should the company modify its corporate policy to allow it to take on the new product?

In the electronics business, it is not uncommon for the most appropriate corporate policy to be the establishment of a new company or a new product division. This is certainly so for the entrepreneurial individual or group.

Allocation of a score against each of the set of criteria makes possible the product profile chart or the derivation of an overall 'figure of merit' for the new product. It can be dangerous to rely too heavily on such formula methods of assessment. Marking schemes of any sort can blur the special characteristics of the subject, and in the quest for product innovation, special characteristics are highly desirable. Therefore, evaluation criteria check lists should be used with moderation. They are probably most valuable in ensuring that all important aspects of the new product are at least thought about. Furthermore, an attempt should be made to identify those factors which are most important to the particular company and to give them appropriate attention. Essential criteria such as adequate profitability must be met. The criteria check list also has the benefits of giving a basis for comparison between different products and projects, of establishing an initial database which can be kept up to date as the product is developed further and of providing an agenda for debate between engineering, marketing, financial and management staff.

Use of a check list of evaluation criteria is only a means to an end and other less mechanistic methods can be adopted. The overall objective of feasibility study is to provide an answer to the basic question: 'can the new product be designed, engineered, manufactured, marketed and sold?'. When enough information has been gathered to formulate an answer with adequate confidence, further pursuit of detail on specific criteria is less important than utilisation of already acquired data – in particular that needed for the business proposal.

The bulk of feasibility study work relates to marketing and technical aspects. The starting point is the product information generated during the concept and definition phase. Production, financial and corporate policy details are either known *a priori* or can be derived from the marketing and technical development results. Fig. 5.2 shows the information pathways linking these components and leading finally to the business proposal.

It has already been pointed out that marketing and technical feasibility studies are best conducted concurrently. They are rather different activities and some comment on what is involved is apposite.

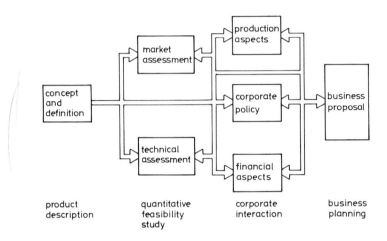

Fig. 5.2.   *From product concept to business proposal via feasibility study*

Market assessment

The assessment of market potential for a new product concept is complicated by the absence of the real product, in contrast to the apparently much more straightforward process of selling actual product to a customer. While the ultimate marketing test is the sales situation, it is clearly sensible to estimate the probability of success in that arena as early as possible, and certainly before substantial sums of money are spent on technical and marketing development.

'Does the market potential justify investment in the new product?'

Addressing this question has been compared to detective work [2]. Techniques include 'desk research', with analysis of published data on markets, and obtaining opinions from people with first-hand experience of the particular market-place. Both approaches have their difficulties; electronic products are often highly innovative and sometimes a market has actually to be created. Published data is always out of date, at least to some extent, and while a specified market sector may be growing rapidly, this is no guarantee that a particular product will get a share. The product is, after all, new, and even highly experienced individuals who may be consulted on marketing will have to consider analogous situations and extrapolate from their knowledge base.

Estimates are first required of:

1 The total size of the market in terms of predicted sales volume each year. For a

new product this may be calculated indirectly by reference to population, industrial and spending statistics and market trends for products of the same type.

2 Market share, or the fraction of the total market likely to be captured.

3 Product lifetime, which has to take account of technological trends and advances as well as market considerations.

4 Based on the above, probability of commercial success in volume terms.

If accurate estimates of these parameters could be obtained, product planning would be greatly simplified. In real life estimation must be supplemented by intelligent guesswork. Market prospects for new consumer products are particularly difficult to predict and information sources to be tapped include appropriate exhibitions and the (carefully controlled) use of marketing consultants with specialised knowledge access. Industrial markets are often more familiar ground for the product innovator, since the product definition is probably formulated originally in response to a specified requirement and since good technical communication is practically assured.

These topics are properly the demesne of the marketing specialists and much study has been devoted to the selection and use of sophisticated market potential assessment techniques. Nevertheless the market assessment task must not simply be handed over to the marketing people; in electronics particularly, and modern technology business in general, engineers must contribute to the marketing work. They play a key role in the preparation of product descriptions and specifications and in interpreting these for the marketing specialists. For industrial products, dialogue with professional colleagues in firms which are potential customers can provide the most reliable market information. Engineering staff also have a responsibility to see that technical possibilities and limitations are not overlooked or underestimated in the evaluation of market potential.

Having established (hopefully) that the new product concept has a worthwhile potential market in volume terms, attention is turned to cost and competition. At this stage optimism usually prevails; it is believed that since the product is attractive, sales will increase rapidly and a high selling price can be achieved, with an associated good profit margin. This optimism should be tempered by caution, in particular as regards prediction of rate of sales increase. Almost always it takes longer to establish a new product in the market-place than was originally anticipated. The activities of existing or potential competitors should be considered; if a new product can be designed and launched on to the market in (for example) 9 months, so can a competitive one. For market potential assessment, realistic selling price estimates are needed. Pricing is itself a specialist topic and there are several well established pricing techniques [5]. Later financial analysis for the business proposal will help to establish the suitability or otherwise of the proposed selling price structure for the new product.

While this market assessment work is being carried out, progress is also being made on the technical front. After perhaps a month, the product's technical specification has become more precise and additional features may have been dreamed up. This information, with schematics, sketches and functional flow charts, can assist the market assessment process.

As the marketing picture becomes clearer, information shortcomings may be evident, and part of the assessment process is to establish if money has to be spent on additional market research, for example by commissioning a limited study of a specific overseas market. The marketing planning process can be initiated at this stage and proposals for test marketing, market launch and product distribution, selling and advertising, with outline costings, can be generated.

This marketing data has to be integrated with technical development information so that meaningful production, financial and corporate policy parameters can be quantified.

Examination of the market potential for the new product may uncover the fact that a similar item is already in production and on sale, perhaps overseas. While this may be upsetting at first, reflection will bring the realisation that an early warning of competition is rather preferable to an anticlimactic market launch. In this situation, consideration should be given to negotiating a licence from the foreign manufacturer for the requisite 'know-how'. This approach can be cost effective and can complement the company's own development programme. Commercial arrangements based on licensing benefit both licensor and licensee; an example is where a US manufacturer has a product which needs 'Europeanisation' to conform to technical or legislated standards. An established European company can provide the necessary technical development, manufacturing and distribution functions rapidly and economically, compared with an approach involving the setting up of a completely new subsidiary. Specialist professional assistance for license negotiations is usually well worth while.

Even if acquisition of existing technology is not a practical proposition, close examination of competitive or related products from both marketing and technical viewpoints can be illuminating. Beware of the 'not invented here' syndrome.

Technical assessment

It is very difficult to determine whether a circuit or system can be designed and built without considering how the work is to be performed in practice. The technical assessment aspect of the feasibility study involves such consideration. For the first time in the new product's life, detailed and practical design and engineering procedures are formulated and examined. This examination and the choices and decisions made as a result lead directly to the definition of technical and financial resource requirements which are key components of a practical business proposal.

Technical assessment objectives are:

1 Verification that the technical development work can be carried out.
2 Definition of design and engineering procedures for product realisation.

It is necessary to establish if the product design requirement can be achieved using available or accessible technology, at an acceptable cost and in a time period which allows the product to reach the market place while it still has a competitive advantage. Leaving cost and time aside for the moment, the probability of successful technical achievement is, *a priori*, usually high, in the sense that electronic technology can generally be characterised as representing solutions in search of problems. If a required logical function or algorithm can be defined it almost certainly can be implemented in digital hardware, software or firmware or a combination of these. Similarly, analogue functions can nearly always be realised electronically to a high degree of accuracy.

The real question is whether technical success can be achieved within the applicable business and technical environment, using the available human, physical, time and money resources. The question is to be answered before the actual technical work is carried out, so a prediction is the best that can be done. The degree of confidence which can be placed in the prediction is dependent on the credibility of those responsible for making the forecast and their credibility is in turn a function of technical ability, professionalism and experience. The worst case is that of the 'first time' individual entrepreneur or fledgling company; these candidates are well advised to take nothing for granted and to double check all assumptions. Second opinions can be sought from qualified sources, including outside consultants. The ability to forecast accurately on technical achievement can be verified and assessed by undertaking a limited but realistic practical pilot design project. This experience will help to bring home the contrast between paper design and physical realisation, including practical aspects such as component sourcing and supply.

Can the product design and engineering tasks be completed in acceptable time at acceptable cost? This is more easily answered if there is an 'in-house' experience base of similar product developments; in this case attention should be concentrated on the more novel details. If, on the other hand, a high proportion of the anticipated design work is unfamiliar territory, to the extent that specific components, circuits and structures cannot readily be envisaged, caution is needed. Initially, engineering staff should spend time, perhaps a few days, examining in detail the product functionality. For all but the simplest products, functional subsections should be identified and allocated to individuals or small groups. Then, a few hours around a table will determine whether a realistic and comprehensive design approach is taking shape. The leader of the feasibility study team is reponsible for identifying grey areas and for structuring investigative work to clarify them.

This introduces the thorny problem of the place of practical work in feasibility study. It is almost inevitable that detailed design and engineering investigations will, at least partly, take the form of practical tests and experiments. This is desirable to eliminate uncertainty. Practical work can, however, be overdone at this stage; engineering enthusiasm and doggedness must be restrained. A logic circuit may be breadboarded, perhaps using wire-wrapped sockets for convenience, to establish a response characteristic or to verify an interfacing function. This is practical and sensible, but the temptation to go further and 'lash up' a complete working prototype of the product can be considerable, and indeed sometimes it is no bad thing that practicality can be demonstrated at this early stage. The dangers are that implementation of a rough and ready prototype can be more expensive and time consuming than was at first thought, so that valuable design resources become tied up, and precise examination of design questions is abandoned in favour of achievement of basic functionality. It is unlikely that the available 'off the shelf' components used for a quick breadboard circuit represent the best and most cost effective choice for a productionised version of the product. A lash up or breadboard circuit can be deceptive; non-technical colleagues may assume that all or most of the design work has been done, whereas only the surface of the real product design task has been scratched. For these reasons, there is a particular need for strong engineering leadership to keep the feasibility study work directed to the primary objectives and practical work must be seen in this context. This is not to underestimate the value of realistic tests and practical circuit evaluations. Electronic product development benefits from the ease with which circuits and software programs can be verified and it is clearly common sense to exploit this advantage, but rather more than a lash up circuit working in a laboratory is needed for proper technical assessment of feasibility; the path to a production version of the product must be mapped out with proper signposting of resource needs.

A particular area where practical experiment is useful is that of interfacing with other manufacturers' products or systems. Much design uncertainty can arise in this context, mainly because the new product's specification, which is at best a preliminary version, is to be made compatible with another 'second hand' specification. The possibilities for marginal design are considerable. Documentation from suppliers, including manufacturers' data on components, is not always exact and its infallibility should never be assumed. Verification of component, circuit and system characteristics by practical test is often well worth while.

With computer-aided design facilities, logic functions can be verified quickly. Microprocessor development systems facilitate checking of particular firmware features. In summary, detailed attention to product specifics at this technical assessment stage can save lots of time and money later. Enough work should be done to clarify what is technically possible and useful for the product's realisation, what practical limitations exist and what design procedures are best for the 'real thing'. Practical implementation work, properly controlled, has its place in this process.

A fundamental feature of feasibility study is limited commitment; the effort, time and money spent on the work is determined by the value placed on acquiring accurate information. If the conclusion from marketing and technical assessments is that further investment in the new product is too risky, the cost of getting to that conclusion must be acceptably low. Several feasibility studies may lead to 'no' decisions, for every one which gives a 'yes'. As always in business, there is a risk of getting it wrong, but a definite decision is preferable to continued dithering. Accordingly, the technical feasibility team should be capable of carrying out its work quickly, preferably in a few weeks. The team is likely to be small, with perhaps one or two full time members, supported by specialists where necessary and with a direct and effective communication link to the marketing group. Market information can cause revision of the original product definition and the technical assessment team must be able to respond to changes. Similarly the discovery of attractive new possibilities in product functionality or performance is of interest and relevance to the marketing assessment group. The technical assessment team leader probably has responsibility for additional projects; this can be helpful in ensuring that product design and engineering proposals are compatible with corporate resources and policy.

To return to the objectives of the technical assessment, the feasibility of carrying out technical development and the required design and engineering procedures are to be determined. These topics are of course interdependent, and definition of technical tasks is often the best way of coming to a judgment on the overall technical practicality of the project.

Examples of technical tasks are:

Digital logic and/or analogue circuit design
Printed circuit board layout
Power supply specification for purchase or design
Microprocessor firmware development

These, and all the other pieces of work that are going to be needed if the product is to be designed and built, should be described in specific terms, with component choices and justifications included. Assuming that the product continues to show promise as market and technical assessments proceed, it becomes necessary to establish that these tasks can actually be carried out. To be practical, an individual and a set of technical resources should be identified for each task. This will help to show up gaps in the ability of the company or organisation to carry out the technical development work; it will also serve to highlight staff or resource overloading or lack of realism in work planning.

In many cases it will be possible to 'put a tick' beside each technical task, confirming that all the design and engineering work needed for the new product is within the scope of available resources. The chances of technical success can therefore be put very high (the possibilities of staff resignations and equipment unavailability cannot be ruled out). This prediction, together with detailed work descriptions, forms part of the overall product feasibility assessment. Where one or more technical tasks require expertise or resources which are not at present available, means of getting the work done must be established. Precedents should be sought. For example, a custom chip of a similar complexity to that needed may already have been developed by an outside specialist supplier. Consideration of the commercial terms to purchase outside services leads to an estimation of the likelihood that the task or tasks can in fact be performed. The estimate is of course most useful if it is a clear yes or no. Putting the chances of successful completion of one essential task at 50% is not very helpful or illuminating. If a particular skill is not available, proposals may be made for the use of outside consultancy or the recruitment of new staff. Either of these approaches is an indication of lowered probability of technical success, since the design and engineering work is not currently all under in-house control. If, in previous projects, such additional skills have been acquired and managed effectively, the chances of success for the new project are of course correspondingly enhanced. If a preferred and practical solution to an anticipated design task is not identifiable as a result of technical assessment work, a proposal should be prepared showing what resources will be needed to develop a viable approach to the task. Obviously the technical way forward is then less sure but the effort needed to remove the uncertainty has been quantified.

The overriding considerations in technical assessment are verification and identification. Every technical task necessary for product implementation must be defined and examined to determine that it can be carried out. Ways of avoiding or overcoming anticipated design and engineering difficulties should be summarised. All this information facilitates the preparation of an outline technical development programme with costings for resource and service utilisation. Recommendations made should be justified.

Given the wide range of electronic product complexity possible and the diversity of companies and organisations engaged in product development, a standardised approach to technical assessment is impossible. Nevertheless, the key topics listed earlier as evaluation criteria for technical development feasibility should each be considered in detail; they are likely to be applicable.

The form in which the results of technical assessment are presented can vary widely, from a short report with appendices on technical details to a substantial set of documentation. The form of the output should be suited to its purpose and individual companies can adopt uniquely preferred formats for presentation. Whatever the format, the output should include the definitive product specification. This engineering document is a principal objective of feasibility study work and is the starting point for the design and engineering project, if authorised. The content of the specification has already been discussed in outline in the previous chapter, and further reference to it will be made in Chapter 7, Project Management.

There is, of course, benefit to be gained from the experience of others and there are specific product sectors where some degree of useful standardisation can be adopted. One such is for products based on microprocessor technology, and many UK businesses have benefited from the Department of Trade & Industry's Microprocessor Application Project, particularly in the areas of awareness and feasibility study. The project was launched in 1978 'to encourage the application of microelectronics in industry', and, among other support arrangements, provides funding for authorised microprocessor consultants to conduct detailed feasibility studies on new electronic products [6]. The 'MAPCON' arrangements have been operated on a wide scale, with over 2500 feasibility studies approved for support in the first three years. The experience gained has led the Department of Trade & Industry to formulate helpful guidelines for the conduct of technical feasibility studies [7] and these are often applicable whether or not the product is the subject of MAPCON grant aid. Other countries have launched initiatives for the encouragement of electronic and information technology based industry and considerable attention has been paid to technical feasibility assessment in the context of industrial project viability.

To conclude this discussion of technical assessment, the following is the contents list for a MAPCON feasibility study relating to a proposed microwave security system. It is indicative of the material which has to be gathered to perform the technical assessment of a new electronic product.

Technical feasibility study

Contents

	Page no.
Objectives	1
Summary of recommendations	2
Introduction	3
Analysis of product requirements	4
(i) Operating parameters	4
(ii) Range	4
(iii) Microwave technology	4
(iv) False alarm performance	5
(v) Power supply	7
(vi) Product shape	7
(vii) Product operating modes	10
Signal processing	13
(i) Microprocessor to implement source modulation	13
(ii) Signal analysis	13
(iii) Alarm condition, decision making	17
System outline	19
(i) Selection of hardware	19
(ii) Choice of microprocessor components	21
(iii) Firmware	21
(iv) Development schedule and product costing	23
System specification	24
Conclusions	28
Appendix 1 Radar Range Equation	30
Appendix 2 Fourier Spectral Analysis of a Man Target	31
Appendix 3 Operating Conditions, Ratings, and Characteristics	32

A feasibility example – Sinar Agritec Ltd.

A good illustration of the importance of both marketing and technical considerations in the new product's transition from product concept and definition, through
feasibility study, to commercial realisation is given by the story of the Agritec
Moisture Computer. This product is a microprocessor-based portable instrument
used to determine the moisture content of whole grains, seeds, crop commodities
and indeed many other substances. The measurement procedure is based on a
capacitance technique with a built in electronic weight balance and automatic
temperature correction. On-board programs (in ROM) provide automatic calibration
for 30 grains, seeds or substances.

The Moisture Computer is the first product from Sinar Agritec Ltd., which was
formed in 1978, in Egham, Surrey, UK, to design and sell 'high technology' instruments for the international agricultural market. The company ascribes the initiative
for the product concept to one of its directors, R.H. Fraval, who visited Indonesia
early in the company's life. He learned that the market for agricultural instruments,
and in particular moisture meters, would grow significantly in the Far East during
the 1980s. In the company's words: 'Further investigation revealed that considerable
innovation in moisture meter design was feasible, and after consultation with
experts of the National Institute of Agricultural Engineering in Silsoe, the Californian Department of Agriculture, the International Rice Research Institute in the
Philippines and other organisations, the product specification for a portable, fully
automatic moisture meter of the capacitance type was established.'

The company then examined the product requirements such as ease of use,
accuracy, speed, portability and price and deliberately sought out new and un-
orthodox ways of achieving innovative and cost effective technical solutions. This
technical feasibility study led to the choice of a 16 bit microprocessor structure
for the product, with low power integrated circuits and an LCD digital display. The
packaging requirements were precisely defined. It was seen that newly established
international regulations were very relevant to the product design and to its marketing.

A business plan for the product's technical and commercial development was
formulated. The project manager for the design and engineering phase was Dutch-
man Michael van der Matten, who had a background of professional experience
in the aerospace industry. The case design was contracted to an outside industrial
designer, John Ryan.

Prior to its market launch, the Moisture Computer won first prize for 'the best
invention incorporating a microprocessor' in the 1980 British Microprocessor
Competition jointly sponsored by the UK National Research Development Corporation (new replaced by the British Technology Group) and the National Computing Centre. The judges were of the opinion that the new product 'was bristling
with invention' and that 'worldwide market prospects were excellent'. The company
stated that it benefited from the competition's requirements for a working model
and full documentation by a specific deadline. 'It required late evenings and early

mornings and very good co-operation between all those involved.' Deadlines are essential in any project!

The product was launched commercially in 1981 and received a good market response, including approval by professional agricultural organisations. A Mark 2 version was developed to incorporate operational enhancements and to provide for the computation of Hectolitre weight, thus extending the sales potential (see Fig. 5.3). Even with careful market and technical feasibility study, design changes

Fig. 5.3. *Mark 2 Moisture Computer from Sinar Agritec Ltd.*

may be required after a product first appears in the real marketplace. Sinar Agritec Ltd. has been sufficiently adaptable to cope with newly discovered commercial needs. Like any product for agricultural application, sales of the Moisture Computer are seasonally variable. Sales have also been made in markets such as the food and pharmaceuticals industries. Production level (in 1984) is building up to 100 units per month.

Sinar Agritec Ltd. has obtained valuable information about further product requirements and commercial opportunities from trading in the agricultural sector through several harvest seasons. The company acknowledges the importance of first-hand market experience in determining product specifications and the value of innovative thinking in the evolution of good designs. 'We have more than enough ideas to continue (developing) new products . . . but first we have to investigate the market potential for each one.'

As with Sinar Agritec's Moisture Computer, every product development project must address the complete set of commercial and technical dimensions of business. This holds true for the feasibility study work.

Corporate interaction

If the early results from market and technical assessment work are encouraging, it becomes necessary to anticipate the consequences of a decision to commit to commercial development of the new product. Company 'factions' which will be involved in the product's commercialisation usually have plenty to say about how the project should be handled. It is best if these inputs can be marshalled at an early stage to ensure that misunderstandings and unwarranted assumptions are avoided. The discussion process between the factions, rather grandly termed 'corporate interaction' here, involves production and financial departments together with top management. One objective is to ensure as far as possible that all relevant technical and commercial opportunities and constraints are examined in good time; if it becomes clear that changes or additions will be needed in the organisation's resources or methods, the feasibility of providing for them must be assessed. The primary objective is of course an overall evaluation of the new product's merits in the context of the actual business environment.

In the particular case of the production department, involvement of professional manufacturing staff well before prototypes are fabricated is often critical. Indeed the production department may be the best place for prototype construction, since adherence to manufacturing standards is thereby assured. That is, of course, assuming that actual prototypes are to be made; it is practicable nowadays to undertake comprehensive and fully documented electronic design work using computer-aided facilities. The result is information for manufacturing which makes possible a direct transition from design to production. To many this is drastically optimistic, but the approach is in fact used by major computer manufacturers. In effect, pilot production replaces prototype fabrication. The critical importance of feasibility study here is obvious.

The more traditional approach based on prototype and pre-production models also demands close co-operation with production staff, and the feasibility of the proposed approach to full product realisation is best examined in consultation with the professionals who will be responsible for the transition from development to production. Specific aspects to be reviewed include choice of components and materials, product assembly procedures, testing and maintenance in the field.

Marketing and technical assessments are of limited value without financial data. Initially this should comprise estimates of development funding needed and predictions of overall market launch budgets, selling costs, material, component, assembly and test costs and expenditure on new plant and equipment (if required). These are 'ball park' estimates, with which to construct a general, commercial picture of the new product. If the figures 'hold together' (for example, compared with those for other products already successfully commercialised), such a positive indication of feasibility helps to justify the formulation of a quantified business proposal.

Even a very rapid calculation of materials, manpower and equipment needed for technical development will indicate whether the project is within the scope of the business organisation. Similarly, the likely expenditures on market launch and selling structures can be estimated quickly. At this stage, these figures are probably in round thousands of pounds or dollars, with time scales in months. A preliminary discussion of financing needs for development and commercialisation should include consideration of funding sources, interest charges and related tax concessions and grant support. Will the new product programme fit into the company? If the first signs of success are visible now, the new product will start to gather support within the organisation. Obviously it is best if this support comes from more than one faction, say, engineering. 'High tech is no passport to business success' [8] – alone.

A preliminary financial assessment of the product's commercial viability is probably the most illuminating procedure that can be carried out within the study of feasibility. The financial analysis need not – indeed at this stage it cannot – be exhaustive, but it provides a common basis for product evaluation by marketing, technical, financial and general management groups. Just as soon as market and technical assessments indicate that the product can be sold, developed and produced, the first set of figures should be put together. The figures must relate to the actual business environment; every new or existing company has to cope with operating overhead costs and wage structures and these can vary widely between firms, but since the immediate objective is a 'ball park' calculation, a number of shortcuts can be taken.

A good basic question is 'Will the new product generate sufficient added value?'. Added value is the difference between the per unit selling price (after any discount) and the corresponding cost of 'bought in' materials and components. The selling price to be adopted cannot be set arbitrarily; it must be acceptable in the market place and the market assessment work should have produced guidelines for customer pricing. For novel products, first estimates of the possible selling price are likely to be optimistic. Experience and knowledge of the market-place can help to ensure that the estimated selling price is as realistic as possible. Pricing is a topic in its own right, and the strategy adopted is likely to be a function of corporate policy. At the feasibility assessment stage 'the price which the market will accept' is often most useful, as it involves market knowledge and gives an indication of the commercial room to manoeuvre. The cost of materials and components for product manufacture is a function of product volume which again relates to marketing.

Vendors' break points for costs are usually at 10, 100, 1000 and 10000 quantities and the likely volume range must be determined. For large volumes on agreed 'call-off' supply contracts, special discounts can be negotiated. These can be very important for profitability but their consideration can be deferred to more detailed planning. At this early stage in the product's life 'catalogue' costs are usually adequate.

To take a simple example, an electronic product (a security system) with an anticipated customer (sales) price of £500 and a materials and components cost of £200 will provide an added value of £300. This will be the only amount of money available to cover wages and overheads and to provide a net profit.

An estimate of per unit manufacturing labour cost can be made by using available information on wages costs and a prediction of employee hours needed to assemble and test the product. Wages costs include social security payments and the applicable hourly or daily rate can be calculated. Due allowance should be made for holidays and inefficiency if yearly or monthly salaries are taken as starting points. For example, a manager may find that each assembler is productive for six 1/2 hours per day, giving an hourly wages cost to the company of £3. If a total of 30 employee-hours is needed to assemble and test each product unit, the manufacturing labour cost is accordingly £90. This, added to the materials and components cost (£200) of the example gives a direct manufacturing cost of £290. Subtracted from the selling price (£500), this in turn gives a 'gross margin per unit' or 'contribution per unit' of £210.

This looks healthy enough, but there are other costs to be considered. Leaving the cost of product development aside for a moment, every business organisation has operating overhead costs, (such as administration, travel and selling) separate from direct manufacturing costs and accordingly sometimes termed indirect. The estimation and allocation of overheads to particular product lines can be complicated. Fortunately, some simplification is often possible.

Every company which has been in business for some time has a history of actual expenditure on overheads. This can be analysed, adjusted for inflation and other factors and used as the basis for estimation of overheads attributable to the new product.

One method is to take a fraction of the total which corresponds to the proportion of the indirect costs considered necessary to support the new product in manufacture. For example, a company may have current overheads costs of £92000 per year. By identifying the staff and support functions which will be required for a production rate of 500 units per year, it is estimated that overhead costs will equal 40% of the existing total, or £36800. The per unit overhead cost is then £73·60.

Alternatively, the new product may be seen to have administrative and support requirements and an indirect cost structure quite similar to a product already in manufacture. Overhead costs may then be considered as proportional to direct manufacturing labour costs. To give example figures, the company with a yearly overheads total of £92000 may have an establishment of 20 direct employees, so that each employee-week worked (based on 46 productive weeks in the year) implies overhead costs of £100. At an effective $6\frac{1}{2}$ hour day, 5 day week, this corresponds to £3·08 per employee-hour or just over the direct hourly wages rate of £3 mentioned earlier. In practice an overheads' total several times the direct wages bill is not uncommon, and of course overheads are very dependent on the company's structure and operations. For one unit of product, the corresponding 30 employee hours or approximately 1 employee-week represent an overheads component of £92·40. If half of the 20 direct employees were to be engaged in manufacture of the new product, the corresponding year's overheads costs attributable to this activity would be £46 000, of course. The 10 direct employees would cope with the production of almost 500 units per year (10 employees multiplied by 32·5 hours per week multiplied by 46 productive weeks, divided by 30 employee-hours per unit equals 498·3 units).

The overheads cost per unit manufactured, however estimated, can be used in the analysis of profitability. Usually the figure taken is derived as above on the assumption of a certain level of production. Clearly, lower production levels make it more difficult to cover fixed overheads. There is always a danger of underestimating overhead costs or assuming too high a level of production, so that the per unit overheads amount is unrealistically low. A new product may well require indirect support not accounted for in existing operations, for example, to provide for training of staff in new techniques and to cover unusually demanding sales efforts to launch a novel product. Overheads will be discussed further in the context of the business proposal.

The overheads estimation can get awkward if such items as bank interest charges on a loan to cover development, initial production and new equipment have to be taken into account. Indeed they will have to be taken into account when a detailed business proposal is formulated, but it should be remembered that at this stage the basic commercial viability of the new product is under examination. Fine detail is less important than establishing if the product can be manufactured and sold profitably. If it can, working out how the development and launch can be financed is justified. On the other hand, if it has only a marginal commercial future, with limited profitability which is vulnerable to market pressures for reduced sales price, material, component and wages cost increases and other factors, further consideration of development and commercialisation costs is irrelevant.

To summarise the financial parameters of the new product in the example, the following set of rounded figures could apply (for each unit of the security system product, at 500 units per year):

	£	
Selling price	500	
Purchases: materials and components (added value 300)	200	
Direct manufacturing labour cost	90	
Total direct manufacturing cost	290	
Gross margin	210	(42% of selling price)
Overheads total (estimate)	93	
Net profit	117	(23% of selling price)

This sort of calculation can be performed rapidly and revised as new information becomes available. In the example, the net profit appears sufficient to justify going further.

It is of course also necessary to establish if the company can afford to develop and launch the product. A first estimate of manpower and material costs for technical development, establishment of production and market launch is already available. This must now be related to the product's capacity to make profit.

In the example, the first assumption of production volume was for a sales figure or turnover of £250 000 per year. The indicated profit is £58 500. This product is attractive if estimated development, production set-up and launch costs are less than say £50 000, since it would appear that all start up costs could be paid for by the first year's operations. In fact, profitability does not ensure cash availability. In a growth situation, profits are likely to be represented by stocks and work in progress and indeed extra cash is likely to be needed to fund expansion of the business based on the new product's success. Accordingly 'cash flow' calculations will be required to show just what is likely to happen to the company's income and expenditure. For the moment it is sufficiently reassuring to see that profits are likely; the separate question of funding the operation has to be addressed in detail later, in the context of a business plan. A prudent management team will look carefully at the phasing of expenditures on development and launch and at the likelihood of the required sales volume being achieved in the first year. Nevertheless, this product has survived market, technical and preliminary financial assessment and looks as if it should be taken seriously. The last hurdle, before a product is considered as a real business proposition, is that of 'corporate compatibility'.

To have got so far, the new product must already have satisfied the general criteria adopted by the business organisation concerned. The feasibility study group will have identified potential hazards and formed judgments on the product's ability to overcome them. As new information is built up, it remains important to check that the product is still going to be compatible with the business. Some parameters are less quantifiable than other commercial factors. Who will lead the development team? What effects will the new product programme have on staff structures, deployment and promotion opportunities? Does the product match the desired 'company image'?

Decision or proposal?

The product concept and definition have been subjected to extensive testing during progress through feasibility study. Technical development requirements have been quantified and the company's capability to provide for them has been verified as far as possible. Product acceptability in the market place has been gauged. Input from production and financial departments has helped to clarify the corporate structures and procedures to be adopted for technical development, establishment of production and market launch. The product's basic commercial viability in terms of profitability has been examined. A precise product specification has been formulated.

The acquired information has been summarised in reports and used to acquaint management with the product's prospects and potential. This degree of corporate interaction may well be sufficient to justify a decision on the new product, perhaps at a management meeting. A commitment to technical and commercial development may be made. The feasibility study is complete.

This approach to 'rounding off' the feasibility study with a set of reports and a management decision on commitment to the new product is often used for smaller projects. Decisions also have to be made on planning and controlling the development project and on new assignments for staff involved in the feasibility work.

A more formal approach to completion of feasibility study is to use the results of the work as the foundation for a comprehensive business plan for the new product. A management commitment to proceed with development may already have been made but formulation of a detailed business proposal is a valuable discipline. For substantial projects, the business proposal is likely to be essential; a management decision to proceed with product development may require ratification by the board of directors of the company, or outside funding agencies may be approached to make available financial support for the new project. Plans for the new product can best be presented, both internally and externally, using the vehicle of the business proposal, the next topic.

References

1 ALBALA, A.: 'Stage approach for the evaluation and selection of R & D projects', *IEEE Trans. Engineering Management*, 1975, **EM-22**, No. 4, p. 153
2 BERRIDGE, A. E.: 'Product Innovation and Development' (Business Books, Stockport, 1977)
3 MATHOT, G. B. M.: 'How to get new products to market quicker', *Long Range Planning*, 1982, **15**, No. 6, p. 20
4 TWISS, B. C.: 'Managing technological innovation' (Longman, 1980, 2nd edn.)
5 GRAY, I.: 'The Engineer in Transition to Management' (IEEE Press, 1979)
6 NORTHCOTT, J. *et al.*: 'Microprocessors in Manufactured Products' (Policy Studies Institute, London, 1980)
7 MAPCON: 'Guidelines for feasibility study grants'. Department of Trade & Industry, Warren Spring Laboratory, March 1982, 3rd edn.
8 ALEXANDER, C. P.: 'The new economy', *Time*, May 30 1983, **121**, No. 22, p. 50

Chapter 6

The business proposal

The several disciplines of product development have been introduced. Financial considerations which featured during the feasibility study now become the focus of attention, as the new product's potential for commercial success is critically evaluated. The preparation and exmination of the business proposal represent an important milestone. Heretofore, commitment in terms of manpower and money was limited, in the knowledge that effort might have to be written off if the product's promise was discovered to be flawed. From now on, expenditures tend to the substantial, as design and engineering work progress – that is, if the business proposal is accepted.

The business proposal milestone is the point at which new 'paying passengers' – the investors – can get on to the product development vehicle. It would be premature to describe the vehicle as a bandwagon, but if it shows some signs of potential success, these will be recognised by the astute business observer. Of course it is rather easier to identify a bandwagon after it has passed and when it is too late to jump aboard. Recognising an oncoming specimen is tricky!

As new characters appear on the project, others leave. Perhaps the technical feasibility study was conducted by a firm of consultants, whose task is complete. These personnel changes are a notable feature of electronic product development. The industry is fast moving and it is often the case that a product 'changes hands' as it proceeds through the development phases. A product concept and definition formulated by a market research group could progress through feasibility study in the hands of technical consultants. The subsequent business proposal could be attractive to a venture capitalist group or a regional development agency, with a consequent change of ownership or a licensing deal. Of course in many instances a new product remains with the same company throughout the development process but the nature of the electronics business allows for such transfers of responsibility during the project.

There are many routes to the business proposal milestone, including those trod by individual entrepreneurs convinced of their products' feasibility, without benefit of professional market research, perhaps. The product concept, definition and feasibility study sequence can always be seen in a product's development, but the form taken is not guaranteed to follow the advice offered in earlier chapters of this book. Nevertheless, whatever the background of the product and its 'owner', whatever the methods adopted to produce a specification and independent of the way in which the product's merits and potential have been seen to be outstanding, it is almost certain that support has to be sought for further development. Support is in the form of finance for the project. The business proposal is to justify provision of support by showing that there will be an acceptable commercial return on investment in development and production.

New product development requires investment and good investment decisions are based on good business information, presented as business plans or proposals. The background to a new product proposal may well be extensive and detailed feasibility study carried out within a company; it could also be the specialist knowledge and experience of an individual entrepreneur committed to setting up his or her own business. The business proposal may relate to a single product, as discussed here, or to developments involving a product range or trading innovation. Where more than one activity is involved, the specific requirements and contributions of individual products and trading functions are concatenated to form an integrated plan.

The business proposal for an engineering investment such as that required for a new electronic product development programme is likely to be assessed by a financier — a banker, a potential investor or an officer of a grant giving body. Even where the proposal is presented internally, to a company's board of directors, financial considerations predominate. Someone is being asked to take a risk — to invest in a project. The speculation may be implemented within a business by the deployment of staff and material resources, which also represents investment. The financial content of the proposal is therefore very important.

The structure of the proposal must be appropriate for the business context. Within a small company, most of the background to a project is likely to be well known by all concerned, so attention should be focused on plans and requirements. At the other extreme, an entrepreneurial group approaching a bank for financial support has to 'start from scratch' and introduce the personalities as well as the project.

To be as comprehensive as possible, the approach adopted for discussion here is akin to application by a small company for financial support from a government industrial development agency or a venture capital investment fund. (In many countries, including the UK, individual investors in venture capital funds qualify for tax relief against their incomes. Fund managers wish to take financial interests in new business projects with promise.) Such bodies are prepared to publicise and discuss the criteria used for proposal assessment. The following recommendations are generally consistent with such guidelines.

First, some general advice. The business proposal document is usually the only evidence available to the decision making body for detailed perusal. It must be credible, realistic, positive and outstanding. The executive or committee considering the proposal has other applications to assess, so mediocrity usually assures rejection and has to be avoided.

The document should be both comprehensive and concise. Its objective has been defined [1] as to provide answers to certain basic questions:

1 Who are you?
2 What do you propose to do?
3 What makes you think you can do it?
4 Who wants it?
5 What will I (the investor) get out of it?
6 How much will it cost?

These questions can of course be rephrased more formally to relate to promoters and staff, the product, technical and marketing capability, commercial viability and funding. The corporate finance director (electronics) of the Midland Bank, UK, has recommended that 20 pages is a good size for a technology-based business proposal. What should those 20 pages contain? Financial institutions ask for:

1 a brief description of the project and its history;
2 a description of the people managing it;
3 a comprehensive and realistic assessment of the market for the product and the special problems involved;
4 details of the amount of money needed and how it will be used;
5 a statement of projected profits, cash flows and balance sheets

Various formats can be adopted for this information, which consists of two main themes – first, material on the proposers and their organisation, the product and its market, and second, financial requirements and projections. The first theme has been discussed in the context of feasibility assessment, so just a few further comments will be made. Financial aspects will be given separate treatment, followed by some advice on proposal presentation to a 'live audience'.

As regards the non-financial part of the proposal, this usually comes first, and first impressions are critical. The proposal document itself should be neat and attractive. The title page should clearly identify the project and the proposers. Next, an overview section should provide a concise distillation of what is unique, interesting and commercially attractive about the new product project. Great care should be taken with the wording of this section, since it sets the tone for the remainder of the proposal. The overview may be followed by a description of the new product, a summary of the results of feasibility study and an outline of the proposed development programme, with time scales. Depending on the intended audience, it is sometimes a good idea to present technical and marketing programmes in the form of bar charts, showing planned progress and resources to be committed.

The planned timing of events is important, since it will dictate the pattern of investment and cash flow. To illustrate, here is a very basic summary of key times in the proposed development of the microwave security system used earlier (see p. 65) as an example of technical feasibility study. This hypothetical product will be featured again to illustrate financial projections.

New product development programme

Microwave security system VMJ

	Timing Week number
Design work start	1

Design
Microprocessor system hardware design complete	5
Microprocessor control firmware developed	6
Microprocessor signal processing firmware developed	10
Microwave circuitry design complete	6

Engineering
NC antenna prototyping complete	6
Microprocessor system PCB prototype	9
Microwave circuit prototype	10
Packaging complete	13

Complete prototype system integrated and packaged	17
Prototype packaged system tests commence	18
Product documentation complete	20

Four pre-production models available	26
Sales literature available	26

Pilot production run – 20 units	30

Sales launch – publicity and test market	34

This information could be summarised even further: 'The first complete prototype system is planned for availability 17 weeks after design and engineering work commences, with pre-production models ready after 26 weeks, pilot production scheduled for week 30 and sales launch at week 34'.

The proposers will have taken care to allow some time for unforeseen delays in this work programme, so that the need to rewrite part of the product firmware in the light of test results (for example) does not render the sales launch date impossible.

This outline of the technical development plan can be conveniently presented as a simple bar chart (Fig. 6.1). It is prudent to check the assessor's preferences in this context and to adopt the most acceptable form of presentation.

The business proposal needs information in summary form and more detailed plans are appropriate to actual project management, which is discussed in the next chapter. To a large extent the development programme for a business proposal represents a skeleton structure, with detail to be added when approval is won. Having defined the work to be performed, it is now necessary to convince the decision makers that the proposers and their staff can make the new product a commercial success. The character and credibility of the 'promoters' are key factors in this assessment. In business, people are more important than projects – the same project can be made into a success or a failure, depending on who is running it. No project will succeed without the right team, and so material in the proposal must emphasise the skills, experience and integrity of the individuals who will undertake the task. This can be supported by an organisation chart with agreed responsibilities, and appendices on qualifications and relevant experience. Other appendices can list facilities and equipment needed for the development, identifying related funding requirements.

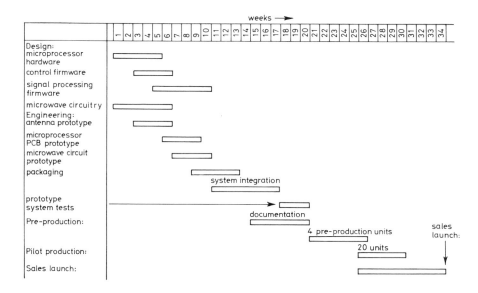

Fig. 6.1.   *Technical development programme: Microwave Security System VMJ*

In the marketing section, attention should be paid to justifying sales projections. Data on market size and competitive products can be checked by the assessors and the proposers' credibility can be enhanced by ensuring that accurate information is presented. It is often helpful to cite sources and authorities for such material.

In general, it must be shown that the project has been thought through completely; good feasibility study can help to ensure that this happens. Attention should be paid to confidentiality; steps should be taken to protect sensitive product and trade information. Material in a business proposal may be seen by consultants engaged to provide a specialist opinion for a bank or government agency. Invariably, permission for this will be sought from the proposers and professional standards of conduct applied, but the risk of disclosure should be considered. Formal assurances and agreements on confidentiality should be sought and obtained. Where it appears that technical invention can be protected by patent law, a patent agent should be consulted. It is also advisable to exclude from the business proposal document, design information which would facilitate product piracy; such material would be of no value to external non-technical assessors in any case. On the other hand, a proposal for in-house consideration may have to include very specific design information, where compatibility with existing manufacturing procedures is involved.

The non-financial part of the proposal should concentrate on informing the reader about the project, emphasising the desirable features. Sometimes a point can be illustrated by an anecdote or a quotation from a recognised authority can reinforce an assertion. Real evidence is valuable and photographs of people, facilities and product details can bring life to a proposal. Remember that the business proposal is a selling document.

The financial plan

The financial part of the proposal is likely to be drawn up with the aid of professional accountancy expertise. What follows is intended to help non-financial people understand what is involved and become able to contribute effectively.

New product development requires financial support. The providers of finance – the investors – wish to make an adequate return on their investment. Hence the questions:

'How much will it cost?'
'What will the investor get out of it?'

These questions have to be answered in the financial section of the business proposal. The conventional set of items in the financial plan for a general business proposal consists of:

Historic accounts
Projected profit and loss accounts
Projected cash flows
Projected balance sheets

Typically, projections for three years are requested, but some bodies are satisfied with two years for certain items.

Straightaway the difficulties of this structure for a financial plan for new product development are evident. Relevant historic accounts are unlikely to exist. Nevertheless they are still asked for by financial people, perhaps because of an overwhelming desire for security – the security of an established business or a fully developed product. To cope with a completely new product, some modification of the list of items is called for, but the proposers should still try to provide most of what is asked for. When figures are presented in a familiar way, the financier is better able to reach a decision.

Help from the computer

Although financial projections have to conform to generally accepted structures and formats which have been used by accountants and businessmen for many years, their preparation can be expedited very considerably with the aid of quite modest modern computer facilities. A personal microcomputer can run the versatile electronic worksheet programs such as VISICALC and SUPERCALC. The success of these management aids is indicated by the hundreds of thousands of copies sold worldwide, and practically every model of microcomputer can support one or both of the programs. Essentially, the user of VISICALC, SUPERCALC or similar program (VISICALC will be used as a generally descriptive term) can build up a worksheet within which arithmetic calculations on rows and columns of figures can be carried out [2]. Any part of the worksheet can be viewed on the computer's display and the complete sheet can be stored on disk. The standard worksheet grid is made up of 254 rows and 64 columns, identified by number and letter co-ordinates. Any grid position can contain data, entered by the operator or computed from other positions' data in accordance with a formula (also determined by the operator). This means that changes to data or formulae can be made conveniently and rapidly, with the calculating drudgery handled by the computer. For financial worksheets, changes such as altered interest or depreciation rates can be accommodated painlessly, and thus 'what if' type operations can be implemented with tentative figures. Details of VISICALC can be obtained from microcomputer manufacturers, computer sales outlets and many computer publications. Users' manuals for these products are of very high quality and the first time operator can be making good use of the program in a very few hours.

For these reasons VISICALC is very popular for generating financial plans, budgets and projections. Accordingly, VISICALC has been used to produce example financial projections (profit and loss, and cash flow) presented and discussed later in this chapter. Before these detailed projections are considered, the background items of historic accounts and development cost require some comment. Development cost has been added to the conventional items already listed for the financial part of the business proposal.

Historic accounts

When a company has been trading for a year or two, audited accounts are normally available. The credibility of the proposers is enhanced if it can be demonstrated that they have the ability to run a successful business, even though the current proposal may not be directly related to that business experience. In brief, if good historic accounts (perhaps for 2 or 3 years) which have some relevance are available, they should be used. In a new product business proposal they could occupy an appendix. In the absence of historic accounts, a statement of expenditure to date on the project (including an itemised breakdown of feasibility study work) can demonstrate that the proposal is based on properly conducted investigation. Furthermore, it can show that the project is in effect underway. Even a little history of achievement can be reassuring. Of course summary financial data on other product successes achieved by the same team should also be included. Normally trading (profit and loss) accounts and balance sheets are needed – these will be discussed shortly in the context of project figures.

Development cost

This is usually unpopular with the financiers since it represents expenditure with no directly related income. It is best to introduce and confront it at this stage. The effects on the complete business programme can be incorporated within the profit and loss, balance sheet and cash flow projections, but it is useful to know what the development costs alone will be. It is very easy to underestimate or underplay the cost of development. This temptation should be avoided – the day of reckoning will come when hidden costs will be revealed. It is better that a realistic estimate which can be adhered to is adopted now. If over-optimism breaks out at this early stage, there will be no resources available to cope with the (inevitable) unforeseen difficulties. Here is an example of a development cost summary for the microwave security system product.

Development costs: microwave security system VMJ: (Direct cost only, overheads not included)

		£
1	Design staff (microwave and microprocessor circuits and systems and documentation)	15 000
2	Outside engineering services (NC antenna fabrication, PCB prototyping)	4 500
3	Industrial design (packaging)	3 500
4	Materials and assembly for prototypes	2 500
5	Preproduction costs (assembly jigs, test equipment)	12 000
6	Pilot production run (20 units)	8 000
7	Market launch, publicity advertising etc.	16 000
		61 500
8	*Less* Government grant for R & D on microprocessor-based products (25% of eligible costs), for example, UK MAPCON arrangements.	8 375
	net	53 125

In this proposal, overheads on development will be accounted for in the overall financial projections. In some instances, grant aid may include a provision for overheads. The grant contribution here is included for completeness; it is usually prudent to disregard likely income from grants until eligibility and approval are formally confirmed. A bird in the hand . . .

Projected profit and loss accounts

The ever-optimistic engineer may feel that use of the term 'profit and loss' is not very positive; after all, the product has survived a thorough review of its feasibility, which included a preliminary check on profitability. However, many assumptions were made at the feasibility study stage, notably that a steady state production and sales situation would apply. In fact the process of reaching a stable and profitable trading position requires time and involves expenditures on technical and marketing development and the establishment of production and sales structures. In this initial part of the product's life (more than a year in most cases) income from sales is likely to be dwarfed by costs, and trading losses are well nigh inevitable. The investor needs to know if this loss making period will be tolerable, taken in conjunction with projections of subsequent profitable times.

A profit and loss account is a summary of the financial ground gained or lost by a business enterprise. When it forms part of a set of financial projections, figures are usually given on a monthly or quarterly basis for the first few years. From these, yearly accounts can be produced. Profit and loss usually refers to a company's performance. It is possible to generate accounts for an individual product or a specific trading activity if sales and costs are analysed so that allocations can be made accurately and realistically. This is what is needed for financial planning for a new product. The special case of a new company 'starting up' with a single new product is in fact quite common in the electronics business, and will be used as an illustrative model.

A fundamental point is that trading success ultimately depends on adequate and profitable sales, whose value is simply given by the price of a unit of product times the number of units sold. Product price and sales volume are interdependent, making price setting, which is 'perhaps the most crucial of all business decisions' [3] a difficult task. The problem is compounded by the pioneering aspects of a truly new product. So what is the right price? A deceptively simple answer is: 'the one which will give the best profit'. Establishing that figure − pricing strategy − is an important business topic [4]. Alternative approaches include 'skimming, penetration and sliding down'. The first involves setting a high price and aiming for the relatively small customer base prepared to pay more for uniqueness, the second means setting a low price and 'going for volume' while sliding down is a combination of first skimming the market and then coming down in price with increased sales volume. Sliding down allows production to be increased in an orderly fashion as sales volume grows and so is favoured for new 'high technology' products, but the opportunity to apply the technique may be denied by such factors as competition. An important consideration is of course 'what the market will bear' and there is no substitute for the gathering of first-hand market intelligence. It is normally easier to cope with the market consequences of a price reduction than with the results of an increase, and this, coupled with the probability of higher manufacturing costs early in the product's life, argues in favour of setting a higher price initially.

A general strategy is to consider first the price the market will bear (including the influence of existing and potential competitors) and the likely trend in the price−volume relationship. A price for initial (limited) production can then be set, with an eye to later reduction when more precise information and additional production capacity becomes available. Entrepreneurs often underestimate the effect on margins of sales commissions and discounts; the real sales value is the amount which will actually be paid to the company by a customer, distributor or agent.

Break-even Chart

At an earlier stage, in the feasibility study, a quick calculation demonstrated that the new product could make a profit. It is now necessary to consider the question of profitability more precisely, en route to profit (and loss) projections. A convenient manual method, much loved by accountants, is based on the 'break-even' chart. This brings together sales price and volume, and fixed and variable costs. Fig. 6.2 shows the principle.

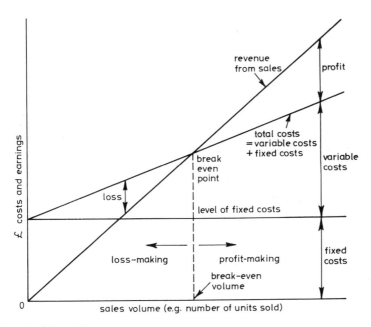

Fig. 6.2. *Break-even chart*

The horizontal axis is for product volume, usually given as number of units. The vertical axis is for money — income and expenditure. For a particular price, sales income (revenue) increases linearly with volume, assuming that the price is fixed. Of course this is not really so, but the chart represents a simplified model of the business.

The costs shown on the chart are of two kinds, variable and fixed. The obvious variable costs are those for materials, components and services used directly in production. The wages for the direct employees who assemble and test the product are also usually treated as variable costs. However, some element of manufacturing wages is almost always fixed in practice since staff cannot be 'hired and fired' to maintain a linear relationship between wages and production volume. For the purposes of the break-even chart, direct materials and manufacturing wages (including social security payments) may be regarded as directly variable with product volume sold. This is a reasonable assumption for a limited production range, say 3000–5000 units of a particular consumer product per year.

Fixed costs include overhead items not directly linked to the level of production. General management, administration and clerical staff salaries, rent and power are examples. Effort should be made to estimate these costs realistically and to allow for increases with time. Setting up for production may well involve capital expenditure on equipment and facilities; the associated maintenance and depreciation costs should be regarded as fixed overhead revenue charges. Financial and legal charges may be difficult to quantify at this stage but some provision should be made initially and revised later.

Some indirect costs, such as advertising, delivery charges and salesmen's commisions, are variable with the level of trading. Where they can clearly be identified as variable, they can be so treated, but in cases of doubt it is better to include an allowance in fixed costs.

With this data, the total costs (the sum of the fixed and variable elements) can be plotted on the chart as a straight line. The intersection of the sales revenue line and the total cost line occurs at the break-even point. Increased number of units sold brings profit. While the chart is based on several simplifications, its accuracy can easily be checked in the important region near break-even by re-examining the figures. For example, quantity discounts on sales or purchases could be desirable or possible at certain levels. The chart may also be used to help determine the selling price necessary at a particular level of production.

The break-even chart is a useful tool in the preparation of profit and loss projections. Its inclusion in a business proposal (perhaps in an appendix) can be informative for the assessors. More complicated models are of course possible and a computer-based version can be manipulated to assess the merit of pricing and purchasing decisions. The basic manual version is normally adequate to illustrate the general price and cost structure of the proposed business. Of course the calculations involved can be implemented readily using VISICALC and displayed graphically using a program such as VISITREND/PLOT. Indeed these same calculations form the basis of the profit and loss projection, since they reflect the relationships between price, cost and volume. Now the dependence of the project on time has to be examined, using figures for sales and costs which have been seen to be consistent with manufacturing capability and profitability.

Calculation of profit or loss for a trading period, manually or using VISICALC, is straightforward. For example, the basic components are as follows (using the microwave security system example):

Profit and loss account: Quarter 2 of year 2

Number of units sold @ £500: 130

		£
Sales Revenue	:	65 000
Cost of sales		
Materials and components	:	26 000
Manufacturing wages	:	11 700
Total direct cost	:	37 700
Gross margin		27 300
Total overheads (revised)		14 740
Net Profit		12 560

In a formal financial plan, profit and loss figures for (probably) three years on a quarterly basis would be required, with detailed breakdown of costs, notably overheads. Expenditures on development which are to be written off should be recorded and accounted for. Of course real assets generated by the development programme, such as PCB artwork and test jigs, have a real value, but this is likely to be a small fraction of the development cost. Accordingly most staff time and consumed materials will appear as expenditures. The value of technical development in an accounting sense is a complicated issue [5]. The safest approach is to view such expenditures as coldly as possible and to write off irrecoverable costs in the year of expenditure. The unsympathetic accountant's approach 'if it cannot be sold it has no value' is likely to be the financier's attitude as well.

So taking the example product (microwave security system) development cost, ignoring possible grant aid and assuming that the assembly jigs and test equipment, together with the pilot production, are worth £7000, the written-off development expenditure is £54 500 over three-quarters of the first year. A two year profit and loss projection prepared using VISICALC could then be as shown in Table 6.1.

Projected cash flows
The profit and loss account does not tell the complete business story. The sequence of production and sale involves delays between transactions and corresponding payments. Purchased materials and components are paid for (usually) several weeks after delivery, and wages and most overheads are paid for in the month the corresponding work is performed. Some time afterwards (weeks or even months) the manufactured products are supplied to customers, who pay their invoices later again. Indebtedness and expenditure precede the provision of credit and subsequent cash income. The sequence is illustrated in Fig. 6.3. Note that the profit and loss account assumes that purchased materials and manufacturing work always have a value, which is increased when product is sold to a customer, thus

generating a profit. From a cash point of view, purchases, wages and overheads are outgoings which are compensated for only when the customer actually pays. At the same time the profit is realised and becomes available to fund further activity. It is generally impossible to expand business as in a start up and growth situation, using only receipts from trading. Outside funding – working capital – is needed to cover the delays in the business sequence. This is in addition to the financing needs of technical and commercial development, and the amounts can be substantial. The cash flow projection is a useful tool in the estimation of total capital needed and the timing of its provision.

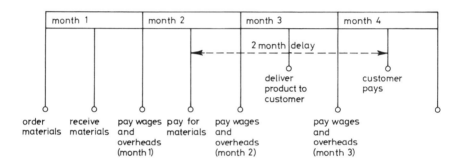

Fig. 6.3.   *Cost, production, sale and income sequence*

A two year simplified cash flow projection showing monthly figures and corresponding to the example profit and loss account of Table 6.1 has been generated using VISICALC and is presented in Table 6.2. The payment delay parameters shown in Fig. 6.3 are used as arbitrary examples. The effect of direct taxes on sales and purchases (such as value added tax – VAT – in the EEC) has not been shown, but is considerable in real projects. The guidance of a professional accountant in tax matters is essential.

The reader is invited to produce, manually or with the aid of a computer, projections based on other assumptions, such as increased time for sales payments receipts or changed levels of production. In practice it is sensible to prepare several cash flows – perhaps a most optimistic scenario, a worst case and an in-between. VISICALC methods are very convenient for such experiments, which are essentially analyses of the project's sensitivity to altered conditions. The process is sure to be illuminating and may lead to strategy changes. At the end the final profit and loss account and cash flow projections should be as realistic and achievable as is possible. The promoters must believe in them and be sure of their ability to keep the promises which are being made in the form of the financial plan.

The cash flows of Table 6.2 assume that £85 000 of investors' funds and a bank overdraft are to be used to cover the cash requirement, with bank interest calculated accordingly. Where the project proposers have made some financing arrangements in the form of their own funds or a loan from some source, such available

Table 6.1 Profit and loss account projection: system VMJ

Profit and loss account projection in £ thousands (£, 000s)

Quarter	Year one				Year two			
	1	2	3	4	1	2	3	4
No. of units sold	0	0	10	40	80	130	120	150
Unit price	0·50	0·50	0·50	0·50	0·50	0·50	0·50	0·50
Sales value	0·00	0·00	5·00	20·00	40·00	65·00	60·00	75·00
Materials/components	0·00	0·00	2·00	8·00	16·00	26·00	25·20	31·50
Manufacturing wages	0·00	0·00	0·90	3·60	7·20	11·70	10·80	13·50
Total direct cost	0·00	0·00	2·90	11·60	23·20	37·70	36·00	45·00
Gross margin	0·00	0·00	2·10	8·40	16·80	27·30	24·00	30·00
Overheads:								
Factory	2·00	2·00	2·00	2·00	2·00	2·00	2·00	2·00
Administration	4·00	4·00	6·00	6·00	6·00	6·50	6·50	6·50
Selling	0·00	0·00	0·25	1·00	2·00	3·25	3·00	3·75
Finance	0·00	0·00	0·01	0·21	0·45	0·39	0·01	0·00
Development:								
Design/engineering	18·50	16·00	4·00	0·00	0·00	0·00	0·00	0·00
Market launch	0·00	2·00	14·00	0·00	0·00	0·00	0·00	0·00
Total indirect cost	24·50	24·00	26·26	9·21	10·45	12·14	11·51	12·25
Profit/loss	−24·50	−24·00	−24·16	−0·81	6·35	15·16	12·49	17·75
Cumulative profit/loss	−24·50	−48·50	−72·66	−73·47	−67·12	−51·96	−39·47	−21·72

For simplicity, no allowance is made for depreciation of capital equipment. Interest is calculated at 1% of each month's bank overdraft figure (the cash cumulative less £85 000 share capital). Materials and components costs increase from £200 to £210 per unit in the third quarter of Year Two.

Table 6.2 Cash flow projection: system VMJ

Cash flow projection in £ thousands (£, 000s)

Year one

Month	1	2	3	4	5	6	7	8	9	10	11	12
No. of units sold	0	0	0	0	0	0	0	4	6	10	15	15
Unit price	0·50	0·50	0·50	0·50	0·50	0·50	0·50	0·50	0·50	0·50	0·50	0·50
Sales invoiced	0·00	0·00	0·00	0·00	0·00	0·00	0·00	2·00	3·00	5·00	7·50	7·50
Cash in (from customers)	0·00	0·00	0·00	0·00	0·00	0·00	0·00	0·00	2·00	3·00	5·00	7·50
Stock purchases payments	0·00	0·00	0·00	0·00	0·00	0·00	0·80	1·20	2·00	3·00	3·00	4·00
Manufacturing wages	0·00	0·00	0·00	0·00	0·00	0·00	0·36	0·54	0·90	1·35	1·35	1·80
Overheads payments:												
Factory	0·66	0·67	0·67	0·66	0·67	0·67	0·66	0·67	0·67	0·66	0·67	0·67
Administration	1·33	1·33	1·34	1·33	1·33	2·00	2·00	2·00	2·00	2·00	2·00	2·00
Selling	0·00	0·00	0·00	0·00	0·00	0·00	0·00	0·10	0·15	0·25	0·38	0·38
Finance												0·22
Development payments:												
Design/engineering	5·00	6·00	7·50	6·00	5·00	5·00	3·00	1·00				
Market launch					1·00	1·00	3·00	6·00	5·00			
Capital expenditure							4·00	3·00				
Cash out (total)	6·99	8·00	9·51	7·99	8·00	8·01	13·82	14·51	10·72	7·26	7·40	9·07
Cash movement	− 6·99	− 8·00	− 9·51	− 7·99	− 8·00	− 8·01	− 13·82	− 14·51	− 8·72	− 4·26	− 2·40	− 1·57
Cash cumulative	− 6·99	− 14·99	− 24·50	− 32·49	− 40·49	− 48·50	− 62·32	− 76·83	− 85·55	− 89·81	− 92·21	− 93·77

Cash flow projection in £ thousands (£, 000s)

Month	1	2	3	4	5	6	7	8	9	10	11	12
						Year two						
No. of units sold	20	25	35	40	45	45	40	40	40	45	45	60
Unit price	0·50	0·50	0·50	0·50	0·50	0·50	0·50	0·50	0·50	0·50	0·50	0·50
Sales invoiced	10·00	12·50	17·50	20·00	22·50	22·50	20·00	20·00	20·00	22·50	22·50	30·00
Cash in (from customers)	7·50	10·00	12·50	17·50	20·00	22·50	22·50	20·00	20·00	20·00	22·50	22·50
Stock purchases payments	5·00	7·00	8·00	9·00	9·00	8·40	8·40	8·40	9·45	9·45	12·60	12·60
Manufacturing wages	2·25	3·15	3·60	4·05	4·05	3·60	3·60	3·60	4·05	4·05	5·40	5·40
Overheads payments:												
Factory	0·66	0·67	0·67	0·66	0·67	0·67	0·66	0·67	0·67	0·66	0·67	0·67
Administration	2·00	2·00	2·00	2·00	2·00	2·50	2·00	2·00	2·50	2·00	2·00	2·50
Selling	0·50	0·63	0·88	1·00	1·13	1·13	1·00	1·00	1·00	1·13	1·13	1·50
Finance						0·84						0·01
Development payments:												
Design/engineering												
Market launch												
Capital expenditure												
Cash out (total)	10·41	13·45	15·15	16·71	16·85	17·14	15·66	15·67	17·67	17·29	21·80	22·68
Cash movement	−2·91	−3·45	−2·65	0·79	3·16	5·37	6·84	4·33	2·33	2·72	0·71	−0.18
Cash cumulative	−96·68	−100·13	−102·77	−101·98	−98·83	−93·46	−86·62	−82·29	−79·96	−77·25	−76·54	−76·72

'Cash cumulative' figure represents funding need — it is assumed that this is provided by £85 000 of share capital from the proposers and a bank overdraft facility.

capital can be shown in the cash flow as reducing (or covering) the cash requirement. For most projects, the financial plan is used to help obtain needed capital.

Of course many new product development projects are funded within existing
companies as part of the overall business operation. In this case the projections are
used to determine the funding needed and the repercussions for other trading and
financial activity can be evaluated. As mentioned earlier the approach adopted
here for business proposal presentation is appropriate to a small company dependent on a single new product. Utilisation of the business proposal, including the
financial projections, to obtain financial support will be discussed shortly.

Before leaving cash flows, it is worth recording that financiers do not consider
them in isolation, despite the fact that they bring together so much information
about the project and its needs. A banker has commented that 'cash flows are
frequently a triumph of optimism over realism'. Accordingly great care should be
taken to ensure realism and this involves presentation of an integrated financial
plan. Profit and loss account, cash flow and the final item – the balance sheet –
must correlate precisely.

Balance sheet projections
The financial progress of a company (in the current context, the anticipated progress) is shown by the profit and loss account and further illuminated by the cash
flow. The balance sheet reveals the overall financial condition of the business on a
given date – usually at the end of a financial year. By reading the balance sheet, a
picture of the firm's health can be derived.

The profit and loss account is a very important component of the balance sheet.
Other accounts which appear include those of the bank, the creditors to whom
money is owed, and the debtors who owe money to the company. The balance
sheet lists these accounts and the amounts in them as either assets or liabilities.
Some 'fixed' asset items are the values of possessions such as equipment and build-
ings. Depreciation reduces the worth of fixed assets, including the tangible results
of development work, and this should be estimated realistically by considering
the useful life of the item and ensuring that the value is written off over that
period. Other assets are stocks of materials, components and finished products.

The 'bottom line' of the balance sheet is the net figure or balance of assets
(positive) and liabilities (negative). This 'net worth' reflects what belongs to the
business after all liabilities have been discharged.

The projected balance sheet must make provision for tax on profits and for
allowances (for example in respect of investment in equipment or development)
against that tax. Tax rules change, particularly where governments are seeking to
encourage particular aspects of business, and professional advice is essential.

Table 6.3 shows illustrative balance sheets for the first two year-ends of the
example product. For increased realism, the figures include an initial share captial
amount of £85 000 (not shown separately in the cash flow of Table 6.2). The
improvement over the second year due to the contribution of trading profit is
evident.

Table 6.3   *Example balance sheet (values rounded to nearest £100)*

		End of Year 1 £	End of Year 2 £
Assets			
(Note 1)	Fixed assets	7 000	7 000
	Current assets:		
	Stocks and work		
	in progress	10 800	30 600
	Debtors (customers)	7 500	30 000
(Note 2)	Bank account	–	8 300
		25 300	75 900
Liabilities			
	Current:		
	Creditors		
	(suppliers)	5 000	12 600
(Note 3)	Bank overdraft	8 800	–
		13 800	12 600
Net assets		11 500	63 300
Share capital		85 000	85 000
Profit and (loss) account		(73 500)	(21 700)
		11 500	63 300

Note 1: Fixed assets figure is estimated value of equipment available after technical development and for simplicity no depreciation is provided for in Year 2.
Note 2: Bank account is in credit at end of Year 2.
Note 3: Funding available is assumed to be £85 000 share capital plus an overdraft facility –
the total to cover 'cash cumulative' amount in Table 6.2.

Assumptions
Usually these will be noted close to the relevant sets of figures. In addition, it is
helpful to summarise all the important assumptions made in the business proposal,
particularly for the financial plan, in a separate section.

The proposal's merit

The importance of new technology-based industry is increasingly acknowledged
and electronics is seen as a premium business. Competent engineering, thorough
marketing, skilled management and good project planning gain valuable points for
the promoters, if they are well presented in the business proposal. The recognition
of worth is very important but it will not outweigh financial shortcomings. Financial objectives are not identical for promoters and assessors but financial growth is

of common interest. Several criteria are in use for the evaluation of projects in terms of financial prospects. They include:

1 Benefit-to-cost ratio, which is defined as the total benefit (usually profit) divided by the total investment, taken over the product's estimated lifetime.
2 Payback period, which is an estimate of the time taken for the cumulative profit to cover the total investment.
3 Return on investment, which is the ratio-expected cumulative profit divided by the product lifetime in years times the investment made.

While these three differ, they can each be used to evaluate one project against another, which is usually what the financier does in the search for a winner. Using the criteria, the project can also be compared with investments of other kinds, such as putting money in a bank deposit account. A worthwhile project has to do rather better than that!

An obvious difficulty with estimating the true value of profits some years hence is the fact that money used for a project investment now would earn interest if retained. To compensate for this, 'discounted cash flow' is used to calculate the 'net present value' of a sum to be earned in the future. The formula is in fact the reciprocal of the standard compound interest formula. The interest rate adopted should correspond to expected borrowing rates. With discounted cash flow, all predicted cash expenditures and incomes can be normalised and totalled to give a total net present value – another basis for project assessment [11].

These evaluation techniques are more likely to be used by the financial assessor than by the project proposer, but some appreciation of them can assist in the preparation of the financial plan and in its presentation.

The business proposal is one of two key exhibits in the trial of the new product project. The other exhibit consists of the promoters, the entrepreneurial individuals who are going to make it all come true. The combination still needs a lot of work to 'sell' the proposal. First, who to go to for support? Second, how to make the best possible presentation of the case?

Sources of finance

In the Western World, many sources of finance for new technology-based projects exist. As a matter of fact, attractive new projects are actively searched for by financial and government bodies, so that investments can be made. There are even instances where technical development is initiated by financiers who have identified a market and technological opportunity. Admittedly there is a preference for developed products, but the international and local business climates are still very favourable to entrepreneurs and product innovators. Despite this, promoters can find the pursuit of funding frustrating. The frustration can be minimised by adopting a planned and considered approach.

The starting point in the campaign can be the actual creation of the new small company. Entrepreneurs who have business experience, perhaps with an existing concern, can probably handle discussions with financial people. On the other hand, the novice businessman with a predominantly technical background may find the unfamiliar terrain confusing. The process of setting up a company introduces the founders to many aspects of business structures. To proceed, consult the friendly accountant. He can put the promoters in touch with a suitable legal adviser and can provide essential advice, including the effects of investments on personal tax situations. The new company can be incorporated with two or more directors, usually the promoters themselves and an initial paid up share capital or equity investment of £1000 or less. Even this modest structure can be of great help, not only in the sense of experience gained but also to give comfort to the financial assessor. He knows he is dealing with a serious proposal when it comes from an organised grouping. One exception is that regional and local industrial development agencies are often happy to help advise on the actual setting up of the new company. An early visit to make introductions and contacts and to avail of such help is a useful prelude to a formal application for support.

With a business structure and a development proposal, the stage is set for progress. Assuming that this new private company is independent of other business organisations (which might have provided funding) the more readily accessible sources of finance are:

self-help;
bank overdraft;
bank loan;
government grants and loans;
venture capital;
combinations of these.

It is impossible to exhaustively describe all of these here, but some principles and features deserve mention.

Self-help
'If the project is good and you really believe in it, why not invest your own (hard earned) money?' The answer is probably that you do not have enough, but even a limited personal commitment to the project can be a major influence with others. Private individuals may be prepared to put something in; social contacts and introductions from the accountant, solicitor or bank manager may open the door to funds. Taxation legislation (in the UK, the Business Start-Up Scheme) often encourages such private investment. Again tax advice is needed. Promoters should be careful that their control of the business is not eroded by welcoming new investors too readily. In the long term, the product is going to be worth a lot and 'giving away' 10% of it now may be a source of friction and regret later. An experienced accountant or legal adviser can provide guidance. Money may be put into the company as payment for shares (giving the investor a voting equity interest) or as

loans on agreed terms. Although the 'self-help' contribution may be but a small fraction of the total requirement (for example £10 000 in £150 000), it can be the corner stone of the project. It brings with it commitment and determination to succeed and can often be used to 'lever' other funding many times its size.

Bank overdraft

Many small companies get started with a combination of promoters' funds, government grants and bank overdraft. Often they continue to depend on the overdraft in later years, even though bankers prefer to regard the facility as temporary credit. Use of an overdraft is attractive because it is (relatively) cheap and very flexible. For a modest project (requiring some tens of thousands of pounds, say) a 'high street' bank may be persuaded to provide overdraft facilities for a significant proportion of the funding. The branch bank manager will interview the promoters and make an application to his superiors. Security will always be sought, almost certainly in the form of a guarantee by the promoters. If things go badly wrong, personal financial well-being is then at risk. The conditions for an overdraft include the credit worthiness of the responsible individuals, so a good record in that department is desirable. In many projects the overdraft plays a role as a smoothing mechanism for cash flow 'bumps', with other funding arrangements to cover 'hard core' borrowing. The cash flow projection provides a schedule for such needs.

Whether the funding is to include an overdraft facility or not (and it usually does), the manager of the branch where the company's bank accounts are kept is a very important person. He should be consulted as early as possible in the project and kept informed. Among other things, the manager is the interface to the many services provided by the banking group. For example leasing of equipment can substantially reduce the capital requirements and should be investigated. Leasing can be very efficient from a company tax point of view.

Bank loan

The bank may prefer to structure a financing package around a fixed term loan, with an effective interest rate higher than that for an overdraft and fixed repayment terms. Such a loan is more appropriate for a business with some history and some assets. The bank will usually wish to take a floating charge on all the assets, and even perhaps on the debtors' accounts, as security.

The clearing banks are not really risk takers in the entrepreneurial sense and accordingly the security demanded by them to support lending may be impossible for the innovative promoters with an unproven project. This has been recognised in the UK and the Department of Trade & Industry can provide 'selected commercial institutions', including the clearing banks, with security in the form of a guarantee for up to 80% (in 1984) of the amount borrowed for approved projects. The business assets have to be pledged (even though they do not cover the guarantee amount) and the loan is more expensive to service than a conventional one. Such a loan guarantee scheme is an encouragement to entrepreneurism and represents one of the many ways that government co-operates with financial institutions in industrial development.

Government grants and loans

In many countries the range of government backed incentives for 'new technology' is considerable. In the USA, specific tax reliefs and tax credits are directed to encourage new product development. In the UK, central and regional schemes abound. Specific arrangements depend on location, with the most attractive grants available in designated development areas. By and large, 25% or more of product development costs can be directly funded, for example under the Microprocessor Application Project (MAP). Other grants are provided to offset costs of capital plant and equipment, market research and training.

The project promoters can find it difficult to get to grips with the various arrangements. They should accept the fact that all government schemes have very well defined rules and regulations and proposals must conform if they are to be accepted. This may mean rewriting part of the proposal or altering the business plan. It is worth while to consult the appropriate government sponsored organisation before a formal application is made, to establish what is involved. In the UK this could be the Department of Trade & Industry or a regional body such as the Scottish Development Agency or the Industrial Development Board for Northern Ireland. There is certain to be red tape, but it can lead to really useful financial support for the project.

Government loans and investments are possible through bodies such as the UK British Technology Group (BTG, incorporating the National Research Development Corporation). BTG's small companies division can invest directly in companies. It has a particular remit in respect of inventions and products originating in universities and polytechnics. In the USA and Europe, governments have supported the establishment of innovation centres to assist technology-based projects. In these centres, special facilities and educational programmes are provided for entrepreneurs and guidance in the quest for grant aid and financial resources is made available [6].

Venture capital

The venture capitalist is a risk taker. Nowadays, with favourable tax legislation, private individuals put money into risk capital investment funds which are managed with the objective of capital growth. Several of these funds concentrate on high growth industry such as electronics. In the UK a long established source of venture capital is TDC (Technical Development Capital), the 'high-technology arm' of the Finance for Industry Group, owned by British banks but raising money entirely in the private sector.

Generally, equity investment groups such as these do more than take a shareholding (always a minority) in a promising young company. They will provide management support (often with a nominee director on the board) and assist with further financial negotiation. Their slogan is 'high risk, high growth' and they expect to pick a sufficient proportion of winning projects to make financial gains. Usually the investment group takes a flexible attitude in setting up a financing deal structure, which may include features such as preference shares and loans.

After five years or more, the group will wish to realise a capital gain by sale of equity.

Venture capital groups are most interested in projects of substance, which require investments in the broad range £50 000 to £1 million (in the UK). The US has, of course, many finance companies specialising in venture capital, including those sponsored by the Small Business Administration.

Financial packages

The sources of finance which have been briefly introduced are important components in funding new product business. A particular project is likely to require a specific combination of investment, grants and loans and this can only be worked out by a series of negotiations. The relationships between equity and borrowings of different kinds have to suit the proposed business. It is a fact that most business proposals require revision when subjected to detailed examination and the real test of implementation. This is particularly true for projects based on new product development, including electronic examples, and a 'second round' of financing is often needed. Clearly every effort should be made to get the proposal right first time, and worst case scenarios should be considered. They have a nasty habit of coming true or being exceeded in the scale of misfortune.

It is a common aspiration to fund further business growth from earnings, after development is completed and sales are well under way. This is well nigh impossible in practice; growth in sales requires additional working capital, as a cash flow will show, unless exceptional levels of profitability are achieved. Also, development and associated costs tend to continue; they are essential to retain competitiveness and inevitable because engineers never know when to stop. So the promoters are likely to continue meeting financial backers. After a few years they will know rather more about the possibilities and practicalities, including arrangements for share flotations and placings when the company has achieved real success.

The presentation

A beautiful song can be sung badly and the unfortunate performer may be rejected by the audience. Business is as tough, or tougher; an excellent proposal presented incompetently can mean rejection of both proposers and proposal. The acceptability of the presenters influences the judgment of a proposal to a sometimes astonishing degree – in many situations promoters' credibility outweighs project specifics. Often, the promoters are given but one opportunity to demonstrate their ability and integrity, at an interview or meeting. The formula for success is take presentations very seriously.

The variety of projects, promoters and assessment situations possible is such that specific guidelines on presentation tactics are not generally applicable. Indeed, the diversity of human nature and character ensures that the same advice on how best to communicate and impress could be quite appropriate for one individual situation and very unhelpful for another. At a presentation, personality and attitude are on display and it is counterproductive to impose any regimentation of approach, but retaining freedom of expression and individualism is not inconsistent with attention to detail and proper homework. Good preparation leads to good presentation. The following is a mixture of comments and recommendations.

Broadly speaking, presenting a proposal happens in one of two ways – at an 'intimate' gathering in the proximity of an important person's desk, or at a meeting involving several or many people. Sensible rehearsal in advance and a style appropriate to the occasion are called for.

For example, two or three entrepreneurs wishing to discuss the financial requirements of their business with the bank manager would be wise to make an appointment in good time. It is helpful to indicate who will be attending and to agree on the probable length of the interview ('If you can let us have an hour of your time, please?'). Numbers present are important. The bank manager, financier or executive who presides at an informal business discussion is most comfortable and confident (and possibly sympathetic) when he is not too outnumbered. To meet with one assessor, a practical limit of three project representatives is realistic. This can be increased if it is known that additional support is likely to be brought in on the assessor's side, perhaps in the form of an assistant. Two person meetings can be satisfactory and good decisions can be made at one-to-one encounters, but the benefits of having one person thinking while a colleague is talking can be considerable. Two project representatives presenting the case to a single assessor is often an efficient arrangement. Whatever the interview structure, the presenters should agree on the division of responsibilities. It is best to have a 'lead' speaker, to outline the proposal and orchestrate the contributions of colleagues.

As regards what is said, it is dangerous to assume that the audience is familiar with the content of the proposal – it may have received only cursory perusal a few minutes before the interview. On the other hand an assessor may be offended (justifiably or not) by reiteration of written material. The proposal may have been read thoroughly, and penetrating questions prepared. The safest approach is to start off with a reaonably concise overview, using words different from those in the written proposal. The project can be described in a precise, conversational manner without excessive technical detail. Use of a sample of the technology, a product mock-up or photographs helps such an introduction. Triggering the real interest of the assessor is a valuable step towards commitment. A few diagrams or sets of figures can be tabled and frequent reference should be made to the written proposal.

Some of the questions raised by the audience can be anticipated, particularly if this is not the first airing of the proposal. Bankers are concerned about security, so available collateral can be listed in advance and outlined when asked for. If a really 'fast ball' is bowled and an agreed, believable answer is unavailable, not much face is lost by undertaking to get a considered response as soon as possible. If discussion between the presenters is needed immediately, a recess can be asked for. This is clearly disruptive, so queries requiring reflection and discussion should be consolidated and tackled together. It must be shown, by demeanour and action, that the project promoters work together as a team. Several people trying to talk at once may be a convincing demonstration of enthusiasm – the first time it happens. Too much of it can be very irritating.

With experience, it is possible to sense when a case is being won or lost at a meeting. If there are indications of a positive response, press home the advantage by discussing the next steps to be taken and agreeing on a course of action. Often this involves obtaining more information for the assessor. Undertake this task willingly and establish that a decision can then be made at the next meeting. Where the assessor is clearly looking for a way to get the promoters out of the office, take the hint, but first get as much value from the experience as possible – find out why the case is unattractive. By appealing to the experience and kindness of the assessor, valuable insight and clues for 'improvement of the act' can be extracted. It is easy to be discouraged but persistence pays off. The chances of success improve with experience and identified inadequacies can be rectified. Be prepared to revise the proposal and its presentation to make it more attractive to the target audience.

Sometimes promoters have to make their case at a large meeting or even as a formal presentation at a seminar or special conference. This approach has been adopted in the USA to bring together entrepreneurs and venture capital resources. Obviously in such circumstances special attention has to be paid to effective communication, with at least a little element of showbusiness.

Again, preparations are important. If possible, visit the venue in advance and check out the facilities, where to stand, where to put the notes. Some utilisation of visual aids is recommended – a controlled use of colour slides, or best of all, transparencies (not too complicated) on an overhead projector. Ensure that the necessary hardware will be in place and working. Two presenters for a single project is an absolute maximum, with one beginning and ending the talk and the other presenting some key information. For a substantial audience (more than 20) it is much preferable to have just one presenter – if others must be seen, they can be introduced briefly. They can of course contribute to the answering of questions after the formal presentation.

Some presenters make the mistake of reading from a prepared script. This is usually dull and disastrous. Speaking 'from the heart' and from an outline which lists key points and topics comes across logically (because it is structured) and lively. The audience has to be won over. It can be wary of too much showmanship. A professional, enthusiastic presentation encourages the belief that the project will be handled in the same way. When the presentation is finished, by all means breathe a sigh of relief, but do not relax. Questions put from the floor or more privately are critical – the questioners are the ones who are interested.

'Selling' the business proposal, whether to a friendly group of company colleagues, an enigmatic financier or a discerning audience of potential investors, is an essential part of the product development process. The promoters have to sell the new product, the project, themselves and their ability to deliver the goods.

References

1 GILMOUR, A. S.: 'Engineering investments – an approach to management education for undergraduate engineers', *IEEE Trans. Engineering Management,* 1976, **EM–23,** No. 4, p. 157
2 LEWIS, C.: 'Managing with Micros' (Basil Blackwell, Oxford, 1983)
3 ICFC CONSULTANTS: 'Pricing, costing and overhead control', *Nat West Small Business Digest,* April 1983, No. 9, p. 2
4 GRAY, I.: 'The Engineer in Transition to Management' (IEEE Press, 1979)
5 'Accounting for Research and Development'. Statement of Standard Accounting Practice, the Institute of Chartered Accountants in England and Wales, Dec. 1977
6 HAYWARD, G. 'Innovation Centres' SEFI Conference on the Education of the Engineer for Innovative and Entrepreneurial Activity, Delft University of Technology, June 1982

Chapter 7

Project management

New product development project work extends from concept to manufacture. Commitment is increased as development progresses. This is most marked at the business proposal stage, when authorisation is given for major expenditure of human and physical resources on design, engineering and preparation for manufacturing and marketing. The project work pattern changes at that point from finding out and reporting activity to making and doing procedures. A matching restructuring of management arrangements is necessary to provide detailed control and monitoring of design, engineering and the transition to production. This is implemented by the establishment of a project team (Fig. 7.1). The team leader may be known as project manager, project director or by some other title, and ought to have a direct line of communication to top management, which has delegated the task of product creation. The project manager's superiors retain the responsibility for integration of the new product with corporate operations and strategy, as part of overall project management.

All aspects and phases of new product development need management but the demands of later stages are greatest. This discussion of project management is relevant to the complete process, but it is focused particularly on the planning and control requirements of design and engineering. It aims to provide some filling and colour for the earlier outlines of project activity, with an emphasis on human factors.

The need to plan

In general human endeavour, planning is an important option. The variety of human nature ensures a wide range of attitudes to planning. Domestic and leisure activities are obviously affected by the degree to which they are planned, but the satisfaction derived by the participants is not overly dependent on precise specification of objectives and detailed scheduling. Sometimes the individual who 'takes life as it comes' gets along quite well, but such a *laissez-faire* approach does not suit the management of technological innovation. It is just not possible to succeed in the business of new product development without adequate planning. If an engineering group is given funds to cover salaries, expenses, components, materials and required services and is left alone to develop products — with no requirement to report on plans, progress or achievement — it is very unlikely that commercially successful products will emerge from the process. This is not to imply that a skilled group of qualified and well motivated engineers will waste time in idle philosophic discussion or random experimentation. It is quite probable that good investigative and inventive work will be done, as is the case within many university PhD engineering research programmes, but the work will not lead to commercial success unless commercial requirements are attended to. This requires planning based on realistic business proposals.

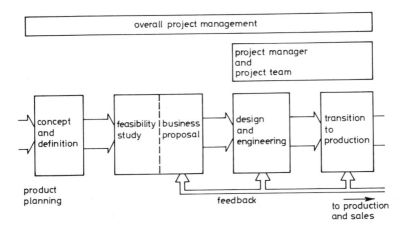

Fig. 7.1.   *Management and project phases*

The term commercial success is used here in its most general sense. Where a company develops and launches its own new product, commercial success can mean that the return on the investment made in technical development, production and marketing is adequate. Where product development is being carried out for a client by an R & D organisation, commercial success can mean that the work is completed to the customer's satisfaction within a budget expenditure which relates acceptably to the project income. In all product development situations, there are close linkages between the achievement of technical objectives, time and cost. These linkages form part of the management network which includes marketing, production and financial control as well as engineering development. The planning process must recognise these business relationships as well as technical criteria. Every new product has a commercial purpose. Its development is not undertaken for the gratification of engineers — although such gratification may be a well justified consequence of project success.

Planning to achieve technical and business objectives can be carried out by face-to-face discussions, jottings on the backs of envelopes, preparation of detailed charts and schedules or utilisation of a computer system. The approach adopted depends upon the nature of the organisation and of the people in it. There is no magic technique for general application. Sound management principles do exist for project planning and the project manager should be aware of them. Furthermore, many helpful procedural approaches to R & D have been published, with itemised procedures and checklists. These can be useful where a fairly narrow and well defined technology is being used – for example, the application of microprocessors to manufacturing in a particular industrial sector, but often marketing, financial and other business dimensions are ignored or glossed over. Attempts to produce a general recipe for planning in new product development soon founder on the diversity and dynamic nature of the activities involved.

In Randall's 1980 report on 'Managing New Products' for the British Institute of Management [1], it is noted that there is a general consensus of those involved in the work that 'neat, orderly schemes found in textbooks on product planning do not fit the untidy reality'. The untidy reality extends through the complete product development sequence. The project manager must plan, but the planning technique adopted, its detail and its flexibility, must match the development task.

Project management principles

Projects must be managed so that development work is directed towards the achievement of the desired commercial and technical objectives and to ensure that the work is controlled to keep it on course. Like all good management guidelines, exhortations such as this seem obvious. Experience shows, however, that the practical application of useful guidelines is not as easy as stating them or indeed believing in their merit. Most design and engineering development projects take longer and cost more than was originally anticipated, sometimes with disastrous commercial consequences for the company concerned. In industry in general, senior management often lacks an adequate appreciation of what is involved in new product development [2].

The electronics business, which has its technology well integrated (usually) with commercial activity, is in a better position to cope with the demands of project management. Nevertheless, electronics projects can and do get out of control. The opportunities for deviation from 'mainline' project work in electronics appear endless. For example, interesting ideas are pursued in preference to getting the product design completed and the design itself cannot be frozen within agreed timescales. Often design deviations are hidden. Not deliberately concealed; it is simply the case that electronic circuit and system design can be so detailed that only the individual engineer responsible really understands what he or she is doing. Electronics projects incorporate innovative change almost by definition. Design engineers' readiness to adapt to new opportunities presented by newly available components and techniques can cause difficulties in finalising specifications. On the other hand, and more importantly, this innovative flexibility is a major asset and resource for product development. It has to be harnessed so that a satisfactory product 'goes out the door' on time and within budget, at the end of the project. This can be achieved by directing some of the intellectual strength of the development team towards effective project planning and control. The ability to innovate and respond to need can be utilised as effectively in project management as in technical design. Fortunately, management of electronics companies and organisations is often in the hands of those who do appreciate the realities of product development, including the frustrations, the satisfaction from achievement and the need for the painstaking application of engineering technique. Furthermore, the technological resources of electronics in this information age can be deployed to expedite project management as well as design and engineering. The smallest electronics company can often be as effective in product development as a large organisation; sometimes it can do even better, with the help of good project management procedures and efficient use of technology.

The first resource for electronic product design is the professional engineer. An individual's talent for design and innovation and his or her capacity for work can be deployed efficiently or inefficiently. Management at corporate, development team and personal levels is needed for the deployment to be efficient. The question 'that is very clever, but is it what is needed?' should be asked frequently, particularly by the individual responsible for the cleverness.

The general principles of management by objectives (and results) are directly applicable to technical project management. The key questions of what must be done, how and when and for how much must be answered. Definition of required achievement and standards, monitoring of progress and implementation of corrective action complete the set of basic and essential procedures.

The basic management process is cyclical. The plan is formulated to meet the organisation's objectives, the resources are marshalled to implement the plan, work is carried out as planned. Monitoring and control of the work leads to the identification and explanation of deviations from the predicted performance, and feedback to the plan is necessary to accommodate corrections for the deviations. This iterative process should lead to the achievement of the original objectives. It can also cope with modifications to the objectives, which sometimes happen in new product development. The currency of management is information, and the management cycle operation is dependent on a management information system. In business, much of the information handled is financial. For new electronic product development, design and engineering information is acquired, processed and applied. Computer techniques can be used for planning and control of development projects, but whether the system is computer based or manual, the requirement is to bring the new product to commercial life as efficiently as possible.

Project management principles and techniques have to be applied throughout the product development process. To establish a reference environment, consider that the design and engineering phase is about to commence. Perhaps a feasibility study has been carried out and a business proposal has been accepted. A technical development budget and outline plan have been approved. Estimates of manpower, components, services and equipment needed have been formulated. Alternatively, experience with other products for the same market has provided many of the answers to questions on viability and requirements, or the new product may be the brain child of an electronics entrepreneur who has burned the midnight oil to establish that his design concepts can work. There are thus many possible background scenarios leading to the start of the design phase, including the commissioning by a client of technical development work within a commercial R & D organisation or an academic institution. Whatever the preliminaries, it is time to start work. Three key steps are involved:

1 Define the objectives.
2 Plan the project work.
3 Control the project work.

Objectives

An enterprise is more likely to succeed when the objectives are defined and known to the participants. In product design and engineering, achievement of the predetermined functional and performance specification is the obvious overall objective and the product specification is the pivot of the development plan.

It is beneficial to formulate a set of specific objectives which are based on the results of feasibility study, the business proposal and the product specification, and which help to determine the responsibilities of the project team members. Even if 'short cuts' have been taken in reaching the design stage (for example, with only a perfunctory assessment of product viability), there is value in preparing a detailed list of objectives. The process may help to compensate for limited investigation by drawing attention to ambiguous requirements or unrealistic assumptions. While the particular form in which project objectives are specified depends on the product, the business and the personnel, some general rules are applicable:

1 Each objective should specify a single result to be achieved, with its target date for completion and limits on expenditure where relevant.
2 It should be as precise and quantitative as possible, so that achievement can be verified and assessed.
3 It should be clearly understandable by those who will be working to achieve it.
4 Responsibility for achievement of objectives should be unambiguous; 'accountability' for results should be specific to individuals.
5 Each objective should be recorded in concise, written form and agreed by those involved.

Three sets of project objectives are normally appropriate:

(a) objectives defined at the corporate or top management level which relate the project to the business strategy of the company or organisation;
(b) technical objectives for the project team, usually defined by the project manager;
(c) work objectives for the individual engineer.

Clearly, the product specification affects all three sets, with technical detail most important at the team and engineer levels. The sets are defined in the above sequence and are closely interrelated for a particular project. Consultation between levels is essential; objectives should not be imposed 'from above'. In a small company or organisation an individual may operate at all three levels. Here are a few example objectives which relate to the development of an electronic security product. They are not a complete representation of all the objectives which a company might use for such a project and technical detail is omitted.

(a) Corporate level
    To develop a microprocessor-based control panel for use with intruder alarm systems manufactured and sold by this company. Productionised prototype units are to be ready for X November 19XX; development expenditure not to exceed £47 000 (outline product functional and physical specification, production schedule and marketing plan attached). Note that an allocation of funds to cover contingencies may be agreed, but not disclosed to the project manager until absolutely necessary for the project.
(b) Project team level
    To develop a prototype microprocessor-based control panel to the functional specification, physical description and applicable standards attached, by Y August 19XX; budgeted expenditure on the work to this stage is £27 000.
    To provide six productionised prototypes and engineering documentation for the product, for marketing and manufacturing approval, by Z October 19XX; budgeted expenditure on this stage is £20 000.
(c) Engineer level
    To design a minimum chip count microprocessor system to provide control, interfacing and memory functions for the product; design and documentation to be completed by U April 19XX (see attached notes on microprocessor families which may be used). (This is just one of several possible design objectives for individual engineers.)

Therefore, objectives should define what has to be done, when and for how much. They set the scene for engineering achievement (in the context of technical development) and are the basis for project planning and task assignment. A clear understanding and acceptance of project objectives and their priorities help the project manager and team to evaluate their work and to make well informed decisions. However, objectives tell only part of the project story; to make progress, design specifications and implementation plans are needed.

Earlier (in Chapter 4, Project Phases) aspects of specifying product requirements in engineering terms were considered. Obviously, the diversity of electronic products for consumer and industrial sectors precludes standardisation of format. To give an illustration of the relationship between specification, objectives and action, part of the development story of a real industrial product will be told.

Hekimian laboratories inc. case study

This is the development of the Pulse Signalling Module, Model 3934, for use with Model 3900 Communications Test System, by Hekimian Laboratories Inc. (HLI), Gaithersburg, Maryland, USA.

Background
HLI is a successful company developing, manufacturing and selling communications systems, test equipment and instrumentation, notably for use by telephone authorities and organisations. It was founded in 1968 and has many technical innovations to its credit, several of its test techniques having been adopted as industry standards. It employs (in 1984) about 260 people, including 30 engineers and technician engineers in product development. Product lifetimes are typically 4 to 6 years, with production runs of several hundred units for each model. The company president, Dr Robert M. Ginnings, adopts a 'market led' strategy for product innovation.

The Model 3934 concept and definition phase covered several months as HLI assessed customers' requirements and considered the integration of the product into its host system, the Model 3900 modular transmission test instrument.

Extract from specification for Model 3934
General: The Model 3934 transmits, receives and measures commonly used telephone pulse signalling systems. It incorporates microprocessor-based signal generation and measurement techniques. Generation of on/off hook supervision, continuous pulsing and keyboard entered repetitive or dialled numbers are combined with measurements of pulses per second and percent break. Operation can be programmed through the front panel keyboard, with programmed values shown on the front panel displays. When equipped with the remote control feature, the Model 3934 can be remotely commanded to perform any of the measurements or functions normally set by the front panel controls. It is designed for use as a plug-in to the Model 3900 Communications Test System (Fig. 7.2).
Specific:
SEND section
Pulse speed range
DC pulsing 5.5 p.p.s. to 23 p.p.s.
AC pulsing 5.5 p.p.s. to 14 p.p.s.
Speed accuracy ± 0·2 p.p.s.

Pulse ratio range
DC pulsing 20% break to 80% break
AC pulsing 30% break to 80% break
Pulse ratio accuracy ± 0·2% break
Frequencies 2600 Hz, 2800 Hz, ± 5 Hz
Send loop resistance selection 0 to 6·3 kΩ in 0·1 kΩ steps
Display − dual 3 digit
Module size − 4 single-width module spaces

Notes: This represents less than 20% of the specification adopted at the beginning of design. This industrial product is sold (mostly to telephone companies) at a price of $3000 to $3500, depending on supplied options. The performance specification is heavily influenced by communication systems standards.

Fig. 7.2.   *Model 3934 Communications Test System Module from Hekimian Laboratories Inc.*

Development sequence
The specification for HLI's Model 3934 communications test system module was complete in mid-October 1980. The project leader was Walter Mack, Senior Staff Engineer, who recalls that the development programme objectives at that time were:

1   to develop an instrument to the frozen specification;
2   to have 20 deliverable units ready for the end of March 1981.

These rather succinct objectives were in fact very appropriate, since the in-company experience of products of this kind was extensive. Walter Mack was responsible for technical development management and made the major contribution to design and engineering. He was assisted by another engineer (assigned full-time) with other support activity summarised as:

	Man-months
Microprocessor firmware	4
Mechanical design	2
Printed circuit board prototyping	4
Technician input	6

(Total development manpower has been estimated as $2\frac{1}{2}$ man-years)
  A brief history of the development is as follows:

mid-October 1980   – specification frozen and design started.
November 1980   – detailed block diagram finalised.
Early January 1981   – schematics for PCB layout ready.
Early February 1981 – first prototype PCBs.
April 20th 1981   – 20 units built, tested and ready for delivery.

Comment
This performance, while regarded as the norm by HLI, is impressive. Many electronics companies fail miserably to achieve development objectives on time and within budget. Small organisations are vulnerable to the effects of inexperience, over-optimism, staff departures, component delivery and service delays and many other factors, in any combination. Even a large electronics company can have a development project jeopardized by a single factor, such as unavailability of a key component. While planning and control of project work cannot guarantee success, the likelihood of objective achievement is enhanced by the formulation of a realistic work programme and sensible anticipation of potential difficulty.

Planning and control

It is in keeping with the spirit of management by objectives to indicate the objective of the remainder of this chapter. This is to provide some guidance on the planning and control of project work to current and prospective developers of electronic products. Much of the advice is based on experience and observation of real situations and on discussions with practitioners and administrators.

Basic essentials are target definition and target achievement. Every project involves several or many targets and in electronic product development the associated tasks are usually the responsibilities of individual engineers. The solo performance of the electronic design engineer is decisive. This is especially the case within small companies or organisations, where the role of project manager is taken by the managing director or department head and the only other team members are the design engineer and (perhaps) a technician. Many excellent electronic products have been developed using such an elemental structure and the archetype of a very few creative engineers carrying through a major electronic development programme has been adopted successfully throughout the industry. Therefore it is reasonable to make a start on guidelines for developers by considering planning and control for the 'one or two man' project.

The extremes of project management technique for the minimal team, as implemented by the project manager, are:

1 Agree with the designers the general characteristics of the product, in sales terms rather than as technological parameters, and leave them to get on with the job. Interrupt only to ask for additional or altered product features, and to enquire when the prototype or production version will be ready.

2 Spend much time and effort formulating a comprehensive plan for the development project, using critical path analysis or some variant of network analysis [3]. This requires a lot of discussion with the design engineer, to try to pin him or her down on promises of dates for particular achievements. Insist that the plan is adhered to and monitor progress with frequent meetings. Accommodate unavoidable changes in the predetermined plan by detailed revision of the programmed network.

In fact, both approaches can work, but they both contain substantial hazards. In the 'near zero planning' option there is a great measure of blind faith in the design engineer. Quite apart from assumptions of integrity, commitment and professional approach, which may well be justified, no effort is made to ensure that the product requirements are accurately understood and interpreted in technical terms. There is no design-to-management feedback — at any rate, not until it is too late. The potential for disaster, as when the key engineer resigns, is considerable.

On the other hand, the 'masterplan' approach can be very inefficient, in that considerable effort is devoted to the generation of a very detailed plan which is based on estimates that are likely to be inaccurate. Consequently, extensive revisions are needed as the truth is revealed, and the plan may end up following the project's progress rather than dictating it. Of course, formalised planning methods based on network analysis are of great value for the management of substantial engineering projects [4] and further reference will be made to them, but basic aids to planning are more useful for project work conducted by small teams.

The valuable attribute of the first planning extreme is freedom to innovate and the second approach includes the essentials of overall direction and attention to resource allocation and time. Effective planning matches the characteristics of the activity in question. Product development can be described as having a capricious nature. There is a need to cope with changed circumstances: the discovery of a better way to implement a logic function; the impossibility of using a particular component due to cost or availability problems; a failure to fit all the electronics into a desirable physical volume. A further complicating factor is the non-uniform rate of progress achieved by most electronic designers. A single firmware routine or a certain analogue function may take several days to complete, while apparently similar design tasks can be 'run off' in a matter of hours. So much depends on the individual's unique blend of competence and experience and the fine detail of the task.

A compromise between autonomy and direction is required. It can be established by sensible use of agreed estimates of start and finish dates for tasks. Inevitably, some of these dates will be incorrect, but taken together they give an overall estimate of project duration or time to achieve a major target such as first operational prototype. Variation of the actual performance from that predicted is quite tolerable when only individual tasks are affected, so long as overall achievement is kept on schedule. Obviously this is impossible if all the estimates for tasks are found to be over-optimistic. In estimating, the project manager's experience can compensate for the effects of the engineer's enthusiasm and confidence. Sometimes these attributes are exchanged and the experienced design engineer can enlighten a less technically aware superior to practical realities. Dialogue and agreement are essential; understanding of objectives and procedures, established before work begins, can ensure that plan changes are adopted harmoniously.

In fact, an element of schizophrenia can help. If, in the first formulation of the plan, there is a little 'double-think', to the effect that it is accepted that some points which are firmly agreed initially may have to be changed later, revisions are not then so traumatic. A distinction has to be made between alterations which affect the pattern of work only, and those which change project objectives, resource needs and timescales significantly. The former type can be handled within the project team, while the latter demand consultation and agreement with corporate management. Happenings such as the introduction of a directly competitive product by another company or the announcement and availability of a new component or material can come as a complete surprise, despite the best efforts in commercial reconnaissance. Nothing is gained by ignoring new factors and slavishly following a partially irrelevant plan, and acceptance of the need to review and revise in a systematic fashion is part of project planning. This must not be used as an excuse to avoid target and task definition in the first place. It has been observed that 'organisations which make plans do better — even if they do not use the actual plans'. The moral is that the discipline of planning uncovers important features of the proposed work and conditions the participants for the tasks. Having made a plan the natural inclination is to use it, of course, and most people are reluctant to write off investment of planning effort, but for the sometimes moving targets of product development, plans which can have their internal structure modified easily and their overall objectives and targets adjusted if absolutely necessary, are best.

These considerations help to formulate guidelines for development project management.

Guidelines for small teams

Here are some recommendations for project management, with the obvious qualification that variations to suit particular conditions are always possible and often
desirable. They are written for the project manager, who may carry much of the
actual design responsibility as well.

Specification
First, review the product specification. If functional and performance requirements
are incomplete or anomalous, set up a meeting with those executives responsible for
defining the product originally and agree a complete specification document.
Consider the development task and the suitability of the assigned team members
for the work. If unsatisfied, negotiate on replacements and redeployments.

Tasks, signing up and estimates
Call a meeting (likely to last for half a day or a whole day) with the project team
to determine technical targets, tasks and work allocations. Remembering that only
a few people (perhaps three or four) are involved, this is likely to be a somewhat
informal gathering, but everyone present should be instructed to take notes. The
project manager takes the official record of the proceedings.

The overall budget and timescale for technical development has probably been
determined already, perhaps with direct input from the project manager. With the
external boundaries defined, the internal structure of the plan is now to be built
up.

The product specification is broken down into a set of distinguishable tasks.
Divisions may be made between digital and analogue, or between hardware and
firmware, or between front end, processing and output, to give some examples.
Allocation of tasks is to some extent obvious, as determined by the skills of the
team members, and is effectively concurrent with the task definition. This represents the 'signing up' procedure, as engineers agree to achieve specific targets.
Attention has to be given to documentation and support for production, with
agreed responsibilities.

The foundation for the plan structure is provided by the set of targets, associated tasks and estimates of time, manpower, materials and services required. While
building the foundation, the project manager and the team can also plant seeds of
destruction, by poor estimation. Greatest hazards are the complete overlooking of
something that has to be done and the underestimation of time and cost for tasks.
Preventative measures include painstaking analysis of design and engineering work
('have we thought of everything?') and keeping enthusiasm and optimism on a tight
rein. Some true horror stories from past projects may be worth retelling. Indeed,
the past history of achievement by team members or their peers provides good
indications of probable performance, if directly comparable tasks can be identified.

It is useful to generate two or three estimates of time, cost and resources for a
particular test task, perhaps making prior consultations outside the organisation, to
get an assessment of consistency in estimating. Try to determine what would

happen if . . . . Some of the anticipated pitfalls and probably a few unanticipated ones will, in fact, occur. Best case and worst case timings and costs for each task should be estimated, and from them, most likely estimates derived. Hopefully, some tasks will be completed within their estimates to compensate for the inevitable overruns. Despite protestations of determined objectivity, engineers are always in danger of underestimating the total time for a job. Of course, realism and sometimes even pessimism come with age, experience and suffering. The project manager considers team members' contributions to planning with such factors in mind.

Plan

The design and engineering content of the project is in the hands of two or perhaps three people, so the technical programme is in fact the work plan for the individual engineers. While the project team is still together at the first meeting, a first draft of the plan should be put together. A bar chart is a convenient representation, with a week as the reference unit of time. An example was given in Chapter 6 (Fig. 6.1). A tidied up version of the plan bar chart can be generated as soon as possible afterwards.

It is always desirable that the plan coverage extends to the end of transition to production. The total time, manpower and cost requirements of sourcing, tooling, test facilities and training, to mention only some of the activities, usually outweigh design work.

To quote an example, a North London manufacturer of personal computer boards, identifying a market slot for small, low cost private automatic branch telephone exchange equipment, put together a fully working prototype system in 28 days. Preproduction and debugging for production took a further 6 months, illustrating the relationship between time for prototyping and time for transfer to production for this telecommunications product.

Formulation of the development plan is a good first achievement for the project team and a good point at which to end the first meeting. Before breaking up, everyone should be clear on what the plan calls for. Work can now start on the first set of tasks.

The project manager also has to define or approve the individual team members' work schedules, including his or her own. Before doing this, the development plan just finalised should be presented to marketing, production and financial managers involved in the new product project. Agreement on performance dates, particularly if there are any changes affecting the business proposal, is essential to avoid disappointment and conflict later. Of course the wise project manager has some resource in hand (3 man-weeks in a 3 man-month project, perhaps). Similarly, the plan and budget should include some provision for investigation of alternative technical approaches, where there is a chance that design will not work out as planned. Always, effort should be put into study of component choice and availability, including second sources.

'Second thoughts' should be invited from the project team a few days after the plan is established, and considered for incorporation into the programme.

The effort put into planning, involving perhaps one or two working days for all those directly involved, sets precedents for recording, monitoring and controlling project progress.

Control
Project monitoring and control depends on the use of two types of plan — the detailed development plan and the individual engineer's work schedule. The latter is likely to be highly individualistic but there are some inescapable requirements. The project manager must know what each engineer is doing on a daily basis, or as near to that as possible. Checking on progress only once a week can allow days at a time to be wasted on misdirected effort. This is not to say that schedules should be changed on a daily basis; necessary revisions are best incorporated in an orderly fashion, perhaps at weekly intervals (except in crisis situations). The defined work schedule for an engineer should not be intimidatingly extensive, as could happen with a list of detailed tasks stretching several months ahead. In practice, detailed tasks can be defined precisely for only a few or several weeks in advance.

A practical procedure can be evolved from these themes, structured on monthly, weekly and daily interactions.

First, the month's work plan is established for the engineer. A brief discussion with the project manager, with reference made to the development plan, is often sufficient to enable the engineer to draw up the schedule personally. The project manager must judge the degree of assistance to be given. The following is a sample:

Portable Data Logger Project

Work schedule: Weeks 10–13, 1983: E. Vernon

Week no.
10: Circuit design for power supply. Breadboard, test and order parts for prototype models (4 off). Parts list.
11: Incorporate power supply and front end circuitry into system design. Start
PCB layout of master-board.
12: Complete PCB artwork (5 days total). Order prototype boards. Document
board test procedures.
13: Firmware to allow alteration of signal input channel allocations and sampling
times. Test using breadboard system. Document new firmware. Prepare
EPROMs for prototype units.

This format is adequate for an experienced engineer, but others may need more detail. This example uses a week-numbering system instead of dates.

The concise schedule, on a single page, can be pinned up and annotated conveniently. It provides an agenda for the daily conversation with the project manager. The manager may well have more than one engineer and more than one project to supervise and such an instant summary of work is helpful.

A fifteen or twenty minute review of progress every Monday or Friday is used to determine short term revisions to the work schedule. A few days before the end of each month the project manager confers with the team members to formulate the next set of work schedules. Outstanding tasks from the current schedules and newly discovered requirements are incorporated.

This may seem trivial, but it ensures that no team member is going down a 'blind alley', and that actual and potential slippages in the development plan are identified and tackled as early as possible. Schedules are revised in an organised fashion, taking the interaction of team members' work into account, and at intervals appropriate to (usually) modest alterations. This approach can be adopted using 6 week or 2 month schedules, and revisions at 2 weekly intervals, or any other preferred format, but the idea of several tiers of project management interaction and control is the valuable principle.

Unforeseen difficulties and delays, such as equipment failures or delivery problems, ensure that progress is never exactly on schedule for any significant period of time, but after 3 weeks of a 4 week work schedule, about three-quarters of the corresponding work should be completed. Good project management involves rearranging tasks when delays occur so that unproductive 'waiting time' is transformed into time spent on valuable documentation, for example. Team members share the responsibility of maintaining progress, and can often make the most practical recommendations to cope with difficulties. Running out of time is one hazard – running out of money is another. Many projects have problems because bookkeeping procedures in the organisation delay the reporting of expenditures, so that the unalert manager is unaware of the true level of spending. The manager should keep an up-to-date record of expenditure committed and ensure that this is to plan. It can be reconciled with the organisation's accounts when these have had time to take actual invoices (which may be presented weeks later) into consideration.

The project manager who contributes to the product design, either directly or by closely supervising the work of one or more engineers, is in a good position to see the implications of progress or lack of it. When an irrecoverable programme slippage has occurred, or is inevitable, the development plan must be revised. It is always tempting to decide on 'pulling out all the stops' and go for the original target date. This may sometimes be possible, using the safety margins of effort already built in to the plan, but when the resource reserved for emergencies has been committed, and progress is still behind schedule, it is more realistic to accept that performance shortcomings on earlier tasks prophesy further difficulties. At best, the original estimates for remaining tasks will be adhered to. It is almost always expecting too much to hope that in addition, outstanding work and necessary corrective action can be completed simultaneously. Nothing is gained by imposing a news blackout; if the team is working long hours to pull back lost time, the organisation's management people should know about it. They should also know what the likely implications for the product programme are and it is the project manager's responsibility to tell them. Many projects have been gravely endangered or brought to disaster by engineers' pride and secrecy. Good management means honest communication as well as honest effort.

Personnel

The 'workaholic' nature of some engineers may result in a group working flat out all the time. This may produce great results and is not to be discouraged, but the steady worker who keeps something in reserve for crisis situations is also to be respected. The unique human characteristics of the members of the team should greatly influence the project manager's approach to supervision. There must be a willingness to sit down and discuss problems, to pour oil on the troubled waters of rivalry situations and to work to establish confidence in the organisation and in the project. Formalised planning and control structures are essential to provide the targets and measurement for achievement that every person needs, but the bottom line is reliance on professionalism to get the work done.

For small companies and organisations, the degree of dependence on a single valuable staff member can be frighteningly large. What happens if such a key person leaves or is unable to work due to illness or accident? Typically, panic and confusion ensue, with a subsequent painful picking up of pieces — if this is even possible. To minimise the risk, the lone engineer must not be required or indeed permitted to work completely independently and unaided on crucial tasks. It is impractical to assign two engineers to the same design task in most real project situations, and such an action could well reduce productivity. Instead, the project manager has to ensure that the risk of design dependence on an individual is kept to acceptable proportions. Design detail must be recorded within a very few days of its generation. The criterion for acceptability is that an appropriately qualified and experienced replacement for a key designer should be able to get design work restarted after a short acclimatisation period (certainly less than a week). To effect such a transfer of responsibility, the project manager's interpretive role is critical.

Creative professionals sometimes resent planning as a form of imposed dictatorship. This is all the more reason for essential planning and control to be organised in a sensitive and obviously common-sense manner. The approach discussed here is applicable to projects (or parts of projects) from a month to six months in duration with a project team of less than about five members. These are not rigid limits and they encompass a large proportion of electronic product development programmes, and there is indeed considerable scope for customised variations on the procedures.

To give just one real example, a project manager working in the American Mid-West prides himself on being the 'ideas man', which is indeed true. His extensive experience and proven ability allows him to develop rapid outline solutions to design problems and to communicate them verbally (or literally on the back of an envelope) to his development engineering staff. He can keep in mind the overall objectives of several projects at once and can accurately assess concepts and designs against these goals. He is also very effective at 'selling' proposals to top management. Little of this talent is committed to paper, since most communication is verbal and lively. Despite this apparent disdain for documentation, his group is very successful both technically and in a business sense. The secret is in the very detailed work schedules and task lists kept by two of his engineers, on filing cards for convenient access. This semi-clandestine planning reinforces the leader's inspiration. It is no doubt fortuitous that this management compensation exists at all and works so well, but the net result is a real level of planning that is at least as valuable as a system with planning centred on the project manager.

A product development parable

Here is a moral table which shows what can happen when a small company's planning and control are inadequate.

There may be an assured market for the new product and finance may be available for its development, but things can still go badly wrong. Disaster can be caused by selling products too enthusiastically, before they actually exist, and by excessive eagerness to get into 'full production'. Enthusiasm and eagerness are virtues, but they can become enemies of planning and control.

A UK citizen spent several years in North America as the sales manager for a major European corporation. His business was equipment for electricity generating plant and he identified a real market need for specific types of power system fault monitoring instrumentation. These products were not commercially available, so he determined to establish a business to exploit the opportunity. Leaving his employment and returning to the UK he made contact with an existing electronics company and a government sponsored source of development finance. A development contract was signed and in less than 9 months a prototype of the first microprocessor-based product was operational.

This prototype, with promises, was used to obtain significant orders from power authorities in three overseas countries. These orders formed the nucleus of an ambitious business plan, which found favour with the government funding organisation. A new independent company was formed and staff were recruited, the majority from the original collaborating electronics company. (Loss of key personnel is one of the hazards of taking on subcontracted development work. Even careful employment contracts are not a real defence.) Factory premises were acquired and while production was started on the first item in the product range, development work was initiated on three further models. Apart from the original business plan, which consisted mainly of financial projections based on anticipated sales, no formal planning was used for project management.

After about a year, the new company had a full order book for the product range; too full in fact, so that several overseas customers were complaining about missed delivery dates. Products which had been delivered and installed were exhibiting multiple faults, in operating conditions not anticipated or investigated by their designers. Key engineering staff were spending long and expensive periods of time overseas, trying to incorporate design changes on site. Their absence from the factory hindered production and testing, since documentation was poor. New products promised to customers before they were even paper designs remained as concepts and speculative fragments of hardware.

The specialised industrial market sector in question, like many others, has an efficient and informal communications network and before long new orders dried up, old orders were cancelled and payments from customers ceased. With a rapidly degenerating financial position and no prospect of improvement, a receiver was appointed by the bank. Later the company was wound up.

The human and financial resources which this company acquired could have been utilised much more efficiently. The first product could have been properly engineered and tested under realistic operating conditions, before it was oversold to customers. Adequate engineering documentation could have been generated so that production was not dependent on designers' presence and knowledge. New product development could have been planned to match design and engineering resource availability. The products could have been developed, announced and marketed in an orderly sequence and to a timely schedule.

Instead, decision making was based on ill defined and idealistic objectives and on excessive optimism initially, on short term expediency later and desperation last. Too little planning and inadequate control characterised this entrepreneurial dash to business failure.

As a postscript, there is indeed some truth in Francis Bacon's adage 'adversity doth best discover virtue'. Survivors of the failed company, sadder, wiser but still enthusiastic, subsequently set up not one but two new business enterprises. These are in microprocessor control systems and in test equipment, and they have prospered.

Techniques

Individual engineers and small teams are responsible for a lot of design and engineering for new electronic products, but many professionals work in relatively large groups. Development of computer products, telecommunications equipment and military systems, for example, generally involves many separate activities. Each requires specialists and a correspondingly large project team is one result. Other pressures which drive engineers 'out of their own little cubicles' are the increasing market consciousness of the electronics business and the accelerating trend from circuits to systems based on complex chips [5]. Co-operative effort is needed to achieve optimum utilisation of development equipment and designers' work stations and organised interaction with marketing and manufacturing is recognised as a key to product quality.

Planning and control of major projects with their bigger teams is correspondingly demanding. Simple bar charts and work plans cover many of the needs, but more sophisticated techniques are valuable when there are dozens of distinct activities which are interrelated and interdependent. These techniques are mainly forms of network analysis, which can be implemented manually or used with the help of a small computer. While recognising their worth and exploiting their virtue, some caution is desirable. Planning and scheduling techniques should be sufficient to ensure that the project programme and its detail are defined, communicated and understood, and be adequate for monitoring and decision making. More paperwork than is really necessary can be irritating and counterproductive.

Indeed, planning can even be overdone. In this thought provoking and cult generating best seller 'The Peter Principle' [6] , Professor Laurence Peter portrays several human realisations of his principle: 'In a hierarchy every employee tends to rise to his level of incompetence'. One such is 'Rigor Cartis' who is characterised by his insistence that every business activity is conducted in strict accordance with the lines and arrows of his proudly displayed charts, regardless of the inefficiencies or losses which result. Beware of the Rigor Cartis syndrome. Planning is not a replacement for work. Planning is to help get work done.

To return to network techniques for project planning, these have many names and PERT (programme evaluation and review technique), network analysis, critical path methods and critical path analysis are all basically the same. The popular UK term is critical path analysis (CPA). The method is applicable to engineering projects with many components and to the handling of several projects simultaneously [3] . As in all planning, CPA operates on the component parts of a project or projects, identified as tasks or activities. Each activity is represented by an arrow, which starts (tail) and finishes (head) at a circle (an 'event' or 'node'). All activities indicated by arrows entering a node must be completed before the start of activities on arrows leaving that node. Each node is numbered and corresponds to a particular time. To prepare a network representation of a project, the activities are listed and the time relationships between them defined. This is done for each activity by determining what other activity (if any) immediately precedes it and what activity (if any) immediately follows it. With this information an 'arrow' diagram can be drawn. Fig. 7.3 is a simple example for part of an electronic data logger development project. In this form, with each node dated, the network is simply a means of illustrating the sequencing and concurrence of activities, and it can be used as an aid to monitoring.

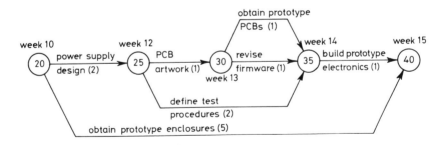

Fig. 7.3. *Arrow diagram*
Estimated task durations are shown in brackets ( ). Each node is numbered and the corresponding event timing is given by the week number

However, much more is possible. The estimated duration of each activity can be examined in the context of all the other activities. The amount of time by which a particular activity's duration can expand without delaying the project is known as 'float'. For some activities the float is zero and the particular sequence or chain of activities which all have zero float is the network's critical path. A project may have more than one critical path and several 'near critical' paths, each with small amounts of float. To shorten the overall project duration estimate, individual critical path activities are examined and the resources allocated to them reconsidered. Changes in duration estimates may change the track of the critical path, with a previously near critical path taking over the role. Inspection of activities with float may lead to economies, for example, by reducing resources allocated to them or by giving outside suppliers of services more time to deliver.

CPA information may be presented as a network diagram or as a tabular listing showing activities with their durations and, for each, definitions of earliest and latest starts and finishes. Start and finish dates (or week numbers, for example) are derived from the 'arrow' network diagram and from the durations. A single entry in the listing could be of the following form, with start and finish week numbers shown:

Activity	Code (tail node number and head node number)	Duration (Weeks)	Start Earliest	Latest	Finish Earliest	Latest
Power supply design	20–25	2	5	6	7	8

In this example there are 3 weeks available (from the earliest start to the latest finish) for a 2 week task, giving a float of 1 week for this 'non-critical' activity.

As a rough guide, network methods including CPA and other variants such as precedence diagrams [7] are useful when a project (or set of projects) involves more than 30 distinct activities. A popular 'conversion' from network to bar chart representation for task scheduling is provided by the 'sequenced Gantt chart' which shows activities as bars drawn to scale. Fig. 7.4 shows part of a sequenced Gantt chart, which is also known as a 'barrow' diagram because of its bars and arrows. As in Fig. 7.3, the arrows indicate activity relationships. Usually all activities are shown as starting at the earliest date and finishing at the earliest finish (using the data in the tabular listing for the network). The amount of float for an activity is shown by shading or cross-hatching the bar from the earliest finish to the latest finish.

Manipulation and analysis of network information to optimise the use of resources and to identify and control critical activities constitutes an important management subject. Many techniques are available, both manual and computer based, to carry out the procedures. Increasingly, software 'planning' packages for microcomputers, based on network methods, are in use, and these can provide analyses of time and resources to correspond to the relationships between activities. Usually a variety of output formats is available, with tabular listings and corresponding bar chart representations of activities. Typically such systems can handle from 500 to 5000 activities, depending on the microcomputer and its peripherals. Such capacity is more than adequate for the clear majority of electronic product developments, but the ease with which plans can be assembled and given a visual representation is very useful. It is clearly sensible to use such a facility when it is readily accessible, for example as one program in the 'electronic worksheet' (VISICALC) repertoire of a small computer.

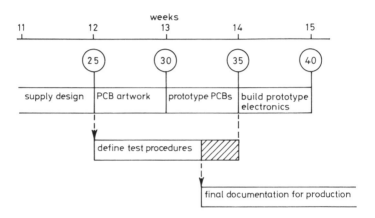

Fig. 7.4. *'Barrow' diagram*
Network node numbers are shown

The increasing use of computer-based work stations for design and engineering work has an analogue in the present day availability of 'management work stations'. Powerful small computer systems with integrated software which includes extensive and imaginative graphics capability can address key business applications, including project planning and scheduling (Fig. 7.5). Their ability to access common files, including those with resource and cost information, and to use this data for both financial and project planning is particularly important. While such systems for business management extend to project planning and control, computer-aided design work stations do not usually provide much help with these tasks. Computer power can, of course, encompass support tasks for project planning and design work simultaneously, and such fully integrated systems exist [8].

Only an indication of techniques and aids for the planning of complex projects has been given. In large companies or organisations, available computer resources and established procedures influence the choice of planning method. Often the overall project plan is generated using a form of network analysis, and this is then converted to bar chart form for scheduling of resources and monitoring. Separate activity groups may have separate and simplified bar charts.

Millimeter wave technology inc. example

This company (MMWT Inc., Atlanta, Georgia, USA) is an example of a successful small business founded on electronic product development. It develops and produces radiometric systems for detecting ice on aircraft, low cost satellite reception antennas for home use and small radar antennas for military and commercial applications. The company has received a Small Business Innovation Research grant from the US National Science Foundation. MMWT's founder and president, Dennis Kozakoff, was able to attract over $200 000 of private investment and a similar amount in loans for the company soon after its incorporation.

Fig. 7.5. *Computer system for management applications*

The business is managed with an emphasis on cash flow control. Long term (two or more years) R & D contracts for government and large corporations are undertaken to smooth out turnover fluctuations. Project planning is based on network analysis, bar charts and work schedules. A small computer is used in several ways: as a design aid, running specialised programs developed by the company; as a financial planning tool, running VISICALC; as a word processor, for documentation, reports and business letters. The computer is regarded as an integral component in business management.

Organisation

The project team should work together for the duration of the design and engineering phase. Where other commitments exist, as for example to production for aspects of an earlier product, team members can be 'loaned' for the necessary work. Careful recording of such time allocations is important in monitoring progress on the development project. The project manager may be responsible for more than one project, and if this is the case, organised control to give proper attention to each is imperative. In some large development establishments, an individual engineer may be project manager for one new product, and may also contribute to design and engineering of another product as a 'rank and file' team member. This flexibility of approach can enhance communications in the organisation.

The combination of management and professional requirements created by new product development and indeed by more general research and development has led to the adoption of a matrix type of organisation in some large companies. Each 'vertical' section, such as production or marketing, has a 'professional' director or head. Projects cut across the sections as 'horizontal' rows or slices, each with a designated project manager and each laying claim to the efforts of individuals from the sections. These team members are responsible to the project manager for the progress of their work on the project for its duration, but are responsible to the appropriate section heads for the professional quality of their specialised work [9].

A very visible aspect of organisation for project management is the meeting. Meetings have been described as 'a concession to deficient organisation' [10] and also as 'nothing less than the medium through which managerial work is performed [11]. There is truth in both descriptions. Meetings are time consuming and therefore expensive. Unless they are clearly useful, many participants will not take them seriously, and so attention has to be given to the proper use of meetings – they are not social gatherings. The importance of regular conversations between the project manager and the team members has already been emphasised and these are valuable meetings. Other meetings, involving more participants, help to disseminate information, build team spirit, solve problems and make decisions.

The objectives of a meeting should be clearly understood by all those attending – in advance, so that needed reports can be circulated in good time and thoughts can be marshalled. Discussion should be initiated, led and controlled by the acting chairman. Here are some 'good management guidelines' for meetings [12].

1 Time is money, so be firm on timekeeping. Specify start and finish times and keep to them.
2 Involve essential participants only.
3 Summarise decisions made before ending, identifying actions needed, when and by whom. Prepare and distribute concise minutes soon afterwards.

Testing and transition

As the product approaches the transition to production, the project team's degree of success in satisfying the requirement specification in a form suitable for manufacturing becomes evident. A crucial criterion is testability, which has to be planned for at the design stage. Microprocessor systems can incorporate self-test routines and less intelligent products can be provided with test points for access by automatic test equipment. It is unfortunate and expensive if a product has to be redesigned after its launch, so that cost effective manufacturing techniques can be applied. In many cases automatic testing is the key to economic production.

Another aspect of testing is that applied at the prototype pre-production and pilot production stages. The golden rule is 'assume nothing'. Even though the first prototype works perfectly (it often does), all other pre-production and early production models should be subjected to as comprehensive a set of tests as possible, including environmental and life testing. A single component change, for example as a result of altered purchasing arrangements, can introduce subtle variations in operating performance. Leave nothing to chance. This is the cue for another true story.

Field trials and test marketing are undertaken to provide trying experiences. They usually do, particularly in the sense that 'if it can go wrong, it will go wrong'. Some years ago a young company developed a computer interfacing system for hospital blood analysis equipment. Since this was before the advent of the microprocessor interface on a chip, the system used specially designed digital circuitry and appropriate machine code software for the host computer. As was so often the case with equipment from different manufacturers, compatibility between the blood analyser and the computer, in terms of signal levels, data formats and timing for data transfers, was very difficult to arrange. In addition to accepting test results, the computer read in patient data on punched cards, concatenated and interpreted results and patient details and produced comprehensive reports. Software was written for these tasks.

During the test phase of the complete system, considerable efforts were made to generate every possible real life situation. The tests were completed satisfactorily and routine use of the system in a city hospital, the first customer, was authorised. The daily work load amounted to up to 18 separate biochemical analyses on each of many hundreds of blood samples, with corresponding patient data. There were some teething troubles, as laboratory staff refined their operating procedures. These were resolved by minor software and hardware changes.

More disconcertingly, there were occasions when sets of results just vanished. Regrettably this tended to happen during visits by important administrators — it seemed that there was a definite correlation between visits and disasters. The sequence of events at such a visit was as follows. First, the consultant biochemist explained the laboratory procedures for receiving and identifying the blood samples and showed how the automated biochemistry analyser performed the various tests. Then he led the visitors to the computer system to await the printout of results.

The emergence of a set of zero values for all tests, for certain unfortunate patients, was not well received. This occurrence was followed by many days and nights of hardware and software probing by the development staff — with no reason for the problem being discovered.

After several such experiences, the relationship between the development team and the laboratory chief had reached a very low ebb. Indeed the team had practically run out of excuses and speculations as to the cause of the fault, and of individuals to present them with any degree of plausibility. The nadir was reached when a group of visitors from another hospital witnessed the generation of more zero results than ever before, with some additional bizarre values. An irate summons was sent to the development company, which fielded the bravest member of the team.

'Allow me to offer you my most sincere congratulations on the excellent work of your staff' was his opening remark to the laboratory chief. 'We have been trying for many weeks to synthesise this situation, but failed. Your expert staff have generated the very conditions necessary to produce the worst case performance. We can now track down the fault with confidence, directness and speed'.

This inspired inventiveness bought 24 hours respite. During this frenzied period it was discovered that certain combinations of punched card data, when read into the computer, caused a register overflow condition which resulted in the corrupted output. A software modification cured the problem and all was well henceforth. The successful resolution of the problem allowed the system to attain credibility, and versions were sold to more than 20 other hospitals.

One moral of this story is that persistence pays off and another is that the ability of the advocate can be crucial, illustrating a further aspect of market consciousness. The principal lesson is that testing should be comprehensive and realistic and should include trials under real operating conditions. While the events narrated predate microprocessor products, the interaction of hardware and software is a shared feature.

Self-defence

The need to guard against deliberate or accidental disclosure of confidential new product information to competitors has been referred to in the context of the business proposal. With the design complete, the risks have increased. Alertness to the dangers of disclosure and reasonable steps to try to ensure confidentiality are inexpensive and often more effective than belated recourse to litigation.

The best defence is to establish a lead over the competition and to hold on to it. Development staff should be reminded of their responsibility to protect their employer's secrets. Corporate loyalty cannot be bought, but a sense of team loyalty is one of the essentials of development work. The project manager can implicitly appeal to the team spirit when he or she identifies specific areas or topics which are not to be talked about outside the organisation. More formally, the employee's contract of employment should specify obligations regarding confidentiality, copyright and ownership of design and patent rights and these can be alluded to at the beginning of the design and engineering phase. Each design drawing and item of documentation should be identified with the organisation's name, the name of the person who generated it and the date. This can help to deter copying for illicit purposes and can provide valuable proof of ownership if copies are in fact made. Discretion is required in discussions with suppliers and subcontractors, who are very probably providers to competitors. They do not need to know about the new product, no matter how enthusiastic its creators are about it. Owner's handbooks and maintenance manuals should be limited to providing needed information – not enough to allow a competitor to copy the product.

Special steps to protect secrets, such as erasing key component identifications in manufacture, can be adopted but it should be realised that a determined and skilled technical pirate can circumvent most obvious defences. When the product is in the market-place, a competitor can buy it, inspect it and take it to pieces. If he has to start development then, from scratch, there is a lot of leeway to be made up. In-house security in development can maximise the leeway, hopefully to a discouraging level.

Legal rights relate to confidential information, copyright, design and patent procedures [13]. A discussion with a legal adviser or patent agent early in the development programme can help to identify formal defence procedures. In general, patent protection should be assessed carefully before embarking on an application. A patent provides a limited monopoly to its owner, for exploitation of an invention. Its practical value for electronic products is often questionable, since product lifetimes are short and technology moves fast, but the cost of patent protection is sure to be considerable. Patent arrangements for foreign countries can be particularly expensive. The 'bottom line' on a patent is its likely commercial value to the project, including the deterrent value. This can only be assessed by investigating the particular case in consultation with a professional patent agent. Registration of a design can be economic and useful, and again the patent agent can advise. A lawyer and a patent agent should be consulted early, and if it is appropriate asked to assist in the construction of product defences. They are much more expensive if they are engaged to rebuild the fortifications after an attack, or to get the invaders outside again. A major difficulty is caused by the contrasting timescales of technological development and the legal process. By the time a case of industrial espionage or infringement of rights brought to law has been expensively resolved, the product's market relevance is very limited. Usually, recourse to litigation on product rights means that the battle is already lost.

Finally, overemphasis on secrecy can be inhibiting. In electronics, good ideas can often transfer from one application area to another without commercial damage. Consciousness of the need for product protection should be tempered by sensitivity to the desirability of professional communication and of remaining up to date with the technology. Both professional and commercial mores should be respected.

The spirit of product development

The next and final chapter is devoted to a case study of the development of a new electronic telephone, so this is the place for a few last words. Experience, observation and the stories told all point to the dynamism and the atmosphere of sweeping change present in electronic product development. In this environment, there are no simple rules which guarantee technical or commercial success, but those with the responsibility of achieving success can improve their chances by the systematic use of information, guidelines, techniques and their own technology.

The spirit of product development is in many ways the team spirit. The individual engineer can make truly impressive contributions to design and innovation, but many other contributions are needed to create a product that can be manufactured, tested and sold. The project manager provides the leadership and expertise to co-ordinate and stimulate the development team, so that a concept becomes a product that works and pays its way.

References

1 RANDALL, G.: 'Managing new products', Management Survey Report No. 47, British Institute of Management, 1980
2 TOPALIAN, A.: 'Design projects are difficult to manage because . . . ', *Design No 345,* Sept. 1977, p. 29
3 LANG, D. W.: 'Critical path analysis' (Hodder and Stoughton, 1977, 2nd edn.)
4 BERRIDGE, A. E.: 'Product innovation and development' (Business Books, Stockport, 1977)
5 ERIKSON A. *et al.*: 'The changing face of engineering', *Electronics,* 1983, **56,** No. 11, p. 125
6 PETER, L. J., and HULL, R.: 'The Peter Principle' (Morrow & Co., 1969)
7 PARKER, R. C., and SABBERWAL, A. J. P.: 'Controlling R & D projects by networks', *R & D Management,* 1971, **1,** No. 3, p. 147
8 YOUNG, R. *et al.*: 'Logic design system manages team projects', *Electronics,* 1983, **56,** No. 10, p. 127
9 TWISS, B. C.: 'Managing technological innovation' (Longman, 1980, 2nd edn.)
10 DRUCKER, P. F.: 'The Effective Executive' (Harper & Row, 1967)
11 GROVE, A. S.: 'How (and why) to run a meeting', *Fortune,* 1983, **108,** No. 1, p. 132
12 LEBOV, M.: 'Practical tools & techniques for managing time' (Executive Enterprises, New York, 1980)
13 JONES, W. S., and PEATTIE, R. C.: 'Micros for managers' (Peter Peregrinus Ltd., 1981)

Chapter 8

Ear to ear and hand to mouth: the story of a new electronic telephone

Ears, hands and mouths have a great deal to do with the performance and acceptability of a telephone. This is the story of the Viscount telephone, the subject of a development programme commissioned by British Telecom (BT) and implemented as a joint development venture by Standard Telephone & Cables plc (STC) and British Telecom Research Laboratories (BTRL). It is presented as a case study to illustrate the planning and management of a major, complex and successful development project for a consumer electronics product. The study is based on information from and discussions with the principal participants: Mr Paul Baker, who was the Project Manager at STC; his colleague Mr Ken Hall, Engineering Project Manager at STC, and Dr Ian Groves, BTRL's Project Manager. The manufacturing view was given by Mr Cyril Brobyn, Manufacturing Project Manager at STC's Northern Ireland factory. Further material is derived from two papers published in British Telecommunication Engineering, the first by Mr C. E. Rowlands of the Product Development Unit, BT Enterprises [1] and the second by Mr R. R. Walker of the Research Department, BT Headquarters [2]. Generous help was given by Mr J. M. Davies, Products Manager, BT Enterprises, Mr D. W. Free, General Manager, STC Information Terminals Division, New Southgate, London, Mr S. C. Curran, General Manager, STC Manufacturing, Monkstown, Northern Ireland and his colleagues Mr F. Sloan and Mr W. Chambers. Viscount product photographs were made available by Mr Evan Kitsell, Senior Product Designer, STC. Production photographs were supplied by Mr Stanley Gilpin, STC Monkstown.

Concept and definition

BT's general business strategy for electronic telephones in 1978 was based on recognition of a need for low cost, high quality products that could be produced in very large quantities and would provide a 'home-base for exports'. With the 'liberalisation' of UK telecommunications in sight, market considerations included the need to make a previously 'invisible' product highly visible to the customer, who would have a choice of models to suit 'personal taste and home fashion'. The product concept was determined by technology push as well as market pull. It had become clear that the evolutionary approach to the use of electronics in telephones (in particular, the substitution of individual electromechanical components by electronic replacements) was providing only a small fraction of the benefits possible from a total electronic design. Advances in LSI had shown that a compact 'all-electronic' design was practicable, with considerable performance, manufacturing and maintenance advantages. BT development staff had already investigated design and engineering aspects of such a product and were well aware of the 'state of the art'. At the same time, STC and other companies were actively involved in the development of specific linear integrated circuits for telephones.

With such general concepts in mind, in mid-1978 BTRL at Martlesham undertook a 6 month study of the technical and commercial advantages of electronic telephones and surveyed relevant sectors of the telecommunications industry in the UK and overseas. The study confirmed that the product concept could be realised, with the involvement of UK industry. Further investigative work was then undertaken, including consultations with STC, to establish design criteria for a product specification and to settle on preferred arrangements for definition, development and production. An important consideration in these deliberations was the need for rapid progress through development; historical timescales for new telephone development programmes were considered to be excessive.

The results of this product planning study were ready in January 1979. They comprised a set of design criteria and a defined strategy for product development.

Principal design criteria

1 The product was to be the simplest member of a possible family of electronic telephones with the lowest achievable cost of ownership.
2 To give improved and more consistent transmission performance (compared with the then current standard telephone).
3 To use a linear microphone, press buttons (keypad) and a tone caller.
4 To exploit the relative freedom from structural constraints afforded by electronic technology to adopt an attractive and consumer conscious case design. The design was to be adaptable for desk or wall mounting.
5 The product was to be designed for automated manufacture.

In general, the adoption of electronic technology was seen as offering considerable benefit in performance, cost of manufacture and maintenance. The standard of construction aimed at was that of 'good quality consumer goods'. The design was to anticipate the later addition of further user facilities (for example multiple number storage and displays).

Product development strategy

A principal objective of the BT strategy was that the sequence of feasibility study, design and engineering and transition to production (to use the terms adopted in this book) should be conducted as smoothly as possible, with minimal delays between phases. This requirement had to be reconciled with commercial prudence and BT decided that the first step was to be the awarding of a 'definition contract' to a selected firm. The definition work was to generate an agreed product specification and to propose comprehensive development and manufacturing plans. A ceiling unit cost to BT for the product was to be agreed.

If the definition contract was carried out satisfactorily, BT's strategy called for the immediate placement of a combined development and production contract, which offered the assurance of a bulk production order on condition that price and timescale targets were met. In August 1979, the project was authorised by BT.

The definition contract

This phase can be seen as a detailed feasibility study and the preparation of a comprehensive business proposal, following the concept and definition phase which established the product requirement and its general feasibility. The definition contract was placed by BT with STC in late September 1979. To obtain the contract, STC had to put a lot of work into formal and informal contacts, and formal proposals and presentations. Specific technical, marketing and commercial statements were prepared and presented. This dialogue was helpful in finalising aspects of the product requirement and meant that effective communication was established between BT and STC. In particular, good engineering communication was assured.

Work on the definition contract took 3 months, with much emphasis on documentation of design proposals and financial analyses. STC staff recall the attention to detail required by BT, with each element of a design proposal subjected to close examination. STC and BT co-operated closely, with the involvement of specialists on maintenance and contracts as well as engineering, marketing, manufacturing and financial staff. Both organisations undertook detailed financial planning, with discounted cash flows and other analyses. Some aspects of this business study were undertaken by engineering staff in the interests of speed and understanding, particularly in the context of manufacturing cost. One comment volunteered on this was that the engineers found manipulation of computer-based financial models easy to learn and to apply.

The definition contract was completed at the end of 1979. The joint development and production plan commissioned by BT was based on four phases, with defined review procedures:

Phase 1: First design phase: completion 46 weeks from start (provision of designs, breadboards, drawings and models)

Phase 2: Development for production: completion 86 weeks from start (multiple prototypes, finalised circuit and case designs, specification of tooling and production equipment)

Phase 3: Qualification and pilot production: completion 125 weeks from start

(models produced at manufacturing plant for formal qualification testing and BT approval, pilot production run of 20 000 units)

Phase 4: Full scale production

On the basis of understanding between BT and STC, the main development programme began in January 1980. Contractual negotiations on the development and production arrangements took several more months, but by mid-1980 administrative details were finalised.

The technical development project

At the beginning of 1980, STC's technical development team for Viscount came together at New Southgate, London. It was led by Paul Baker as Project Manager and Ken Hall as Engineering Project Manager. The collaboration of these two engineers was key to the project's success. Over the $2\frac{1}{2}$ years of the development project, the team had 14 members for most of the time. There were four groups; electronic, mechanical, acoustic and industrial design, plus the continuous involvement of manufacturing which was represented by Cyril Brobyn, Manufacturing Project Manager. Right from the start of design and engineering, the objective of a smooth and efficient transition to high volume production was given high priority. BT's engineering commitment was maintained by Dr Ian Groves, the BTRL Project Manager. Dr Groves' own team had 3 to 5 members, depending on the stage reached in the development programme.

A rough calculation of STC and BTRL technical development man years gives a total figure of around 40. This is one measure of the scale of effort needed for the technical development of a new electronic consumer product with a planned production volume of hundreds of thousands per year.

The starting point for design and engineering work was the very detailed specification of technical requirements and cost objectives which constituted a principal output of the definition contract. This facilitated the formulation of specific task assignments for the STC team and the integration of these responsibilities into a comprehensive project plan.

Project planning and control procedures are described by Paul Baker as 'structured, but in some aspects bordering on the informal'. Certainly communication between and interaction of design groups contained healthy informality, with minimal demarcation and the direct involvement of Paul Baker and Ken Hall, but it is clear from planning documentation that the structured approach was comprehensive and logical. The necessary level of planning flexibility was provided by the use of a form of sequenced bar chart for overall activity scheduling. This was the responsibility of Ken Hall, who also chaired the monthly engineering progress meetings. A part of one edition of the overall activity chart prepared in January 1981, a year into the programme, is shown as Fig. 8.1. It gives a good illustration of how even a multi-task development project can be represented succinctly and

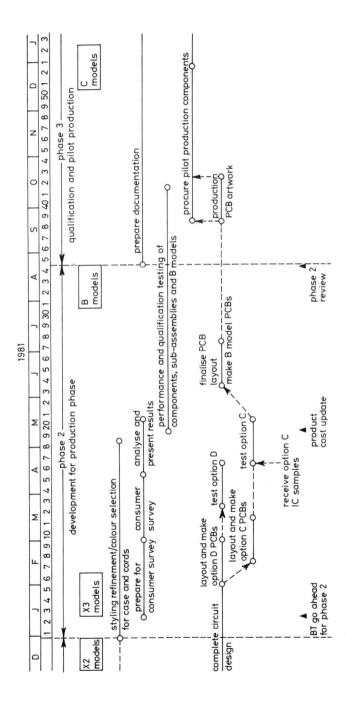

Fig. 8.1. *Overall activity planning chart for Viscount development*

precisely. The overall chart was complemented by a set of eight more detailed bar charts, one for each major activity:

1 Transmission circuit/PCB
2 Signalling circuit
3 Major and minor mouldings
4 Tone caller
5 Keypad
6 Hookswitch
7 Development/qualification test
8 Manufacturing/test facilities

A sample bar chart is shown in Fig. 8.2 for the electronic design task of transmission circuit and PCB.

These planning aids suited the scale and complexity of the project. More formal network methods could have been used but Paul Baker and Ken Hall felt that these could have become 'too rigid'. The importance of achieving goals and the precise definition of goals and milestones were seen as the key factors in planning and control. At BTRL, Dr Groves used his own detailed bar charts to aid development monitoring and control for BT.

The structure adopted for meetings balanced formality with effective communication. It included STC meetings on engineering progress chaired by Ken Hall every 3 weeks, formal technical progress and contract meetings between STC and BTRL chaired by Dr Groves every 2 to 3 months and formal major design reviews chaired by Paul Baker at 6 month intervals. Other meetings covered liaison with marketing, contract staff and manufacturing and quality assurance. Paul Baker gave particular attention to the manufacturing dimension, which involved increasingly detailed meetings as the project progressed.

Work proceeded to schedule and the Viscount was ready for production in April 1982. Of course, there were problems during development. At one stage the critical path was determined by progress on the keypad. Difficulties with materials were overcome by special efforts and some inspiration from STC engineers. Handset design involved several iterations with specialist contributions on acoustics from BTRL staff. On just a few occasions it became necessary to 'declare a slip' in an aspect of the development programme and such instances were compensated for by good planning and extra work.

The final version of the telephone and aspects of its design and construction are shown in Figs. 8.3 to 8.5. Mechanical and electronic design features are presented and discussed in Walker's paper [2]. Only a brief summary of technical innovation in the product is appropriate to this project management study.

Mechanical

Since the product's appearance was not dictated by the electronic technology used, which had modest volume needs, the design team had considerable freedom to optimise the case design for customer appeal and ergonomic considerations. The

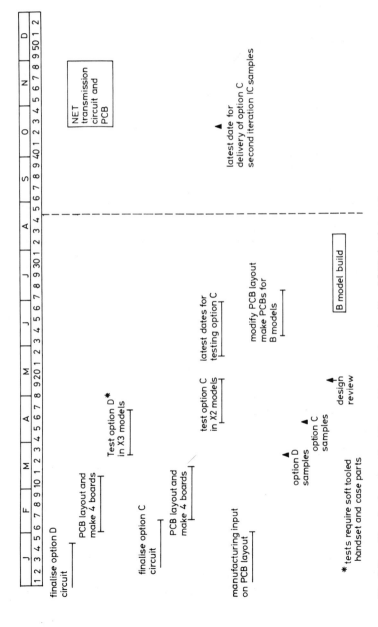

Fig. 8.2. *Planning bar chart for Viscount transmission circuit and PCB development*

final choice of case uses four main mouldings; two body halves and two handset halves, in acrylonitrile/butadiene/styrene (ABS). The keypad has only four component parts and incorporates a conductive rubber key mat. The buttons are smaller than traditional types.

Fig. 8.3.   *The Viscount telephone*

All the mechanical parts are designed for speed and simplicity of assembly.

Electronic
A single bipolar integrated circuit, developed by Standard Telecommunication Laboratories (STL106) is used for speech transmit and receive functions. This chip also incorporates multifrequency signal interfacing for that version of the telephone. Other functions provided by integrated circuits are tone calling, loop disconnect signalling and multifrequency signalling (one or other of the last two, depending on the model). Circuits to combat radio frequency interference and to protect against line voltage surges incorporate discrete components.

All the circuitry is accommodated on a single PCB whose size, shape and mounting are heavily influenced by the chosen case design. This single PCB is cracked across a line during manufacturing, to fit the case (Fig. 8.5).

Transition to production

The new telephone was designed and engineered for automated production and the establishment of the manufacturing facilities kept pace with technical development.

In all, well over £3 million was spent in setting up production, which was initiated in May 1982. The unit manufacturing cost achieved was within the original (1979) STC quotation.

Fig. 8.4.   *Viscount internal structure*

Early difficulties were mechanical rather than electronic, partly as a result of several late detail changes. Many man years were devoted to documentation prior to production launch and formalised 'engineering change orders' coped with the modifications.

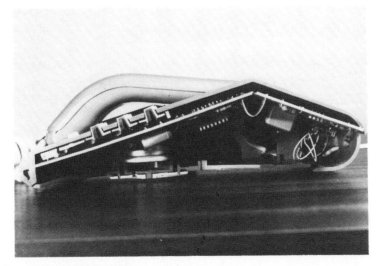

Fig. 8.5.   *Viscount PCB mounting*

Initial minor problems were overcome and production grew rapidly to a rate of over 500 000 units a year. In fact, rates of over 20 000 a week, equivalent to over 1 million annually, have been achieved in 1984 (Fig. 8.6).

Fig. 8.6.   *Viscount production and test at STC Monkstown*

As usual, the transition to production has required attention from members of the development team, despite their new commitments to product derivatives with an extensive range of additional features and functions. Qualification testing of early production output involved considerable effort. Paul Baker believes that such demands can best be coped with by retaining the services of at least one member of each key development group for production assistance, through to an extended point several months after production launch. BTRL's commitment to Viscount has also extended beyond the transition to production. For example, Dr Groves spent considerable time on the telephone, answering technical queries from the most distant bulk purchaser, the New Zealand telecommunications authority.

Final comments

Paul Baker, Ken Hall and Ian Groves each had management responsibilities in the Viscount development programme. Their shared perception of the project emphasises

Fig. 8.7.   *Consumer advertising for Viscount*

the importance of honest and accurate communication within the development team and between engineering, marketing, manufacturing and commercial groupings. Paul Baker adds that communication needs the reinforcement of authority to ensure the implementation of decisions. Ken Hall underlines the significance of communication skills and experience in engineering management. Ian Groves asserts that a successful project integrates marketing and manufacturing with design and engineering from the very start, and it is surely comforting to the engineer taking on project management to hear from Paul Baker that it is important 'if you can really help with technical development problems'.

The final arbiter is the retail customer and the Viscount telephone achieves the intended high level of consumer appeal (Fig. 8.7). The product is a technical and commercial success, nationally and internationally. Its derivatives, with extended functionality, show promise of substantial sales.

The story of the Viscount illustrates the electronic product development process, with its several dimensions and phases and its cast of professionals.

References

1 ROWLANDS, C. E.: 'The electronic telephone', *British Telecommunications Engineering,* Oct 1982, **1**, p. 148
2 WALKER, R. R.: 'Viscount; a new basic telephone', *British Telecommunications Engineering,* Oct 1982, **1**, p. 161

Index

Academic Enterprise Competition, 33
accountant, 79, 93
added value, 69
advocate, 124
Agritec Moisture Computer, 66–68
Apple, 5
assembly, 40
assets, 85, 90

Baker, Paul, 127, 130, 132, 136–138
balance sheet, 79, 81, 90–91
bandwagon, 74
bank
    interest, 92
    loan, 93–94
    manager, 94, 97
    overdraft, 86, 93–94
bar chart, 77–78, 112, 130–132
Bell Holmdel Laboratories, 12
Bell System Picturephone, 11–12
benefit to cost ratio, 92
blood analysis, 123–124
bookkeeping, 114
bottom line, 90
Braun, Professor Ernest, 17
breadboard, 47, 49, 61–62
break-even chart, 83–84
breakpoint, 70
British Microprocessor Competition, 66
British Technology Group, 33, 66, 95
British Telecom, 127–132, 136–137
Brobyn, Cyril, 127, 130
Brown, Professor Wayne S., 19
Burns and Stalker, 26–27
Business Start-Up Scheme, 93

CADMAT, 9, 41–44, 48–49, 62
Californian Department of Agriculture, 66

capital, 90
Carnegie-Mellon University, 33
cash flow, 14–15, 25, 72, 79, 81, 85–89,
    96
    discounted, 92
Center for Entrepreneurial Development, 33
commercialisation, 6, 24–25, 32, 68–72
commercial phase, 21–22
communication, 7, 27, 50
Computing Services Association, 16
concept and definition, 25, 32–36
confidentiality, 78–79, 124–125
contract of employment, 116, 124
controlling, 10
corporate
    compatibility, 72
    interaction, 68–72
    policy, 57
creditors, 90–91
critical path analysis, 109, 118–120
customer, 11, 13–14
custom integrated circuit, 43–44

Data General Eagle, 12–13
debtors, 90–91
definition contract, 129–130
Department of Trade and Industry, 16,
    64, 94–95
depreciation, 90–91
design
    aids, 41
    and engineering, 25, 29–30, 32, 39–50
    approach, 46
    development systems, 41–48
    factors, 45
    productivity, 41
desk research, 58
development agency, 93, 95

directing, 10
directors, 93
documentation, 38, 46, 48, 51, 64, 75−79,
    114−115, 124, 135
Drucker, Peter, 6, 21

Eclipse MV/8000, 12
enclosure, 49−50
engineering change order, 51, 135
entrepreneur, 18, 29, 33, 61, 74−75, 83,
    92−93, 97, 104
entrepreneurial, 3, 29
estimates, 110−112, 118−119

feasibility team, 37−38, 62−63
Finance for Industry Group, 95
financial
    analysis, 38, 69
    package, 96
    plan, 79−90
financing, 69, 75
firmware, 42, 48
fixed costs, 83−84
float, 119
Fravel, R.H., 66
funding, 69

Gantt chart, 119−120
Ginnings, Dr Robert M., 106
grants, 95
Greenberg, Barry, 43
gross margin, 70, 85, 87
Groves, Dr Ian, 127, 130, 132, 136−137

Hall, Ken, 127, 130, 132, 136−137
Hekimian Laboratories Inc., 106−108
Hewlett-Packard, 43
historic accounts, 79−81

Industrial Development Board for Northern
    Ireland, 95
innovation, 2−3, 5, 33, 103
innovation centres, 33, 95
innovative function, 21
integrated circuits, 39−44
interfacing, 62
International Rice Research Institute, the
    Philippines, 66
investment, 50, 54, 62, 75−76, 92−96
investors, 74−76

Jobs, Steven, 5

Kidder, Tracy, 12
Kozakoff, Dennis, 121

Lawrence and Lorsch, 27
leasing, 94
legal advice, 93, 125
liabilities, 90
licensing, 60
LISA, 5
LSI, 41−43, 128

Mack, Walter, 107−108
management
    by objectives and results, 11, 103−106
    consultants, 17−18
    loop, 10−11
    mechanistic, 27−28
    organic systems, 26−28
MAP, 95
MAPCON, 64−65, 81
market
    assessment, 58−60, 68−69
    estimates, 58−59
    planning, 60
    share, 59
Markkula, A.C., 5
matrix organisation, 122
van der Matten, Michael, 66
meetings, 111, 122, 132
microelectronics, 39−43
microprocessor, 39−44, 48, 62, 64
Midland Bank, 76
Millimeter Wave Technology Inc., 120−121
money, 6−7, 14−15

National Computing Centre, 66
National Institute of Agricultural
    Engineering, Silsoe, 66
National Science Foundation, 121
network analysis, 109, 117−120

objectives, 10−11, 104−106
organising, 10, 121−122
overheads, 70−71, 84−89

patent, 125
payback period, 92
PERT, 118−120
Peter Principle, 118
PCB
    layout, 9, 44
    prototype, 47

utilisation, 41–42
pirate, 125
planning techniques, 117–120
power supplies, 49
presentation, 96–99
pricing, 59, 69
product
champion, 7, 29
check list, 54–57
consumer, 59
definition, 35–37
industrial, 59
launch, 23–25
life cycle, 23–25
life time, 9, 41, 59
objectives, 46
planning, 21–25, 59
profile chart, 54–55
screen, 33–34
search, 33–34
specification, 45–46, 64, 73, 105, 111
production, 50–51, 56, 68–72
pilot, 51, 68
profitability, 70–73, 83, 96
profit and loss, 79, 81–85, 90
project
control, 113–114
manager, 29, 46–47, 100, 111–116, 121–122, 126
nature, 25–28
plan, 47, 112–113
team, 46–48, 61, 100, 105, 112–116, 121–123
promoters, 78, 86, 91–98
prototype, 47–49, 61–62, 68, 123

Randall, G., 102
resources, 30, 103, 114, 119
return on investment, 92
Rigor Cartis, 118
Rock-a-Bye Baby Inc., 43
Ryan, John, 66

Scottish Development Agency, 95
security, 94, 97, 125
selling price, 69, 82–83
setting up a company, 93–96
share capital, 90, 93, 95
showbusiness, 98
signing up, 111
Sinar Agritec Ltd., 66–68
Sinclair, Clive, 5
slippage, 114
small business, 15
Small Business Administration, 96
Souder, Professor William, 17, 26
staffing, 10, 28–30, 46–47, 51, 64, 74, 114–116
standards, 49
Standard Telecommunication Laboratories, 134
Standard Telephones & Cables, 127–138
SUPERCALC, 80

targets, 108–110
task dominant, 26
taxes, 86, 90, 93–95
team spirit, 28, 126
technical
assessment, 60–65, 68–69
tasks, 63
Technical Development Capital, 95
technological
environment, 39–44
gatekeeper, 30
technology push, 39
testing, 122–124
Texas Instruments' Speak and Spell, 13
transition to production, 25, 32, 50–51, 122–123, 134–136

UK telecommunications, 127–128
University of Salford, 33
University of Southampton, 33
University of Utah Research Park, 19

variable costs, 83–84
venture capital, 74–75, 95–96
viability, 69–72
Viscount telephone, 127–138
VISICALC, 80, 84–86, 120–121
VISITREND/PLOT, 84
visual aids, 98
VLSI, 9, 41, 43–44
wages costs, 70
West, Tom, 13
working capital, 86
work
plan, 112–114
station, 49, 120
Wozniak, Stephen, 5

ZX81, 5